THE MAKING OF A
PHILOSOPHER

THE MAKING OF A
PHILOSOPHER·

MY JOURNEY THROUGH
TWENTIETH-CENTURY PHILOSOPHY

COLIN McGINN

HarperCollins*Publishers*

FIRST EDITION

Designed by Jackie McKee

Library of Congress Cataloging-in-Publication Data

McGinn, Colin.
    The making of a philosopher: my journey through twentieth-century philosophy / Colin McGinn.—1st ed.
        p. cm.
    ISBN 0-06-019792-7
    1. Philosophy, Modern—20th century. I.Title.

B804 .M355 2001
192—dc21   2001039502

02 03 04 05 06 ❖ RRD 10 9 8 7 6 5 4 3 2

Acknowledgments

I WOULD LIKE ESPECIALLY TO THANK MY AGENT, SUSAN RABINER, for encouraging me to write this book, and for her many useful suggestions. Cathy Mortenson and Galen Strawson were also pleasantly helpful. But my greatest debt is obviously to my teachers, colleagues, and pupils over a thirty-year period of philosophical activity.

Contents

Preface

THE PURPOSE OF THIS BOOK IS TO EXPLAIN PHILOSOPHY IN AN accessible, engaging way. But how best to do that? After trying out a number of plans for such a book, I hit upon the autobiographical format. More orthodox formats inevitably became too textbooklike, and while there is a place for such books I didn't want my book to remind the reader of school. I realized that what was missing was a sense of philosophy as a lived subject—as part of a flesh-and-blood human life. By placing philosophy in an autobiographical context I could convey the excitement and passion of the subject, as well as its trials. I myself have greatly enjoyed reading other philosophical autobiographies and biographies; I might mention here, in particular, Bertrand Russell's *Autobiography* and Ray Monk's biography *Ludwig Wittgenstein: The Duty of Genius*. However much of a purist I may be about philosophy, there is no denying the drama and realism that come with a more personal approach. In this book, therefore, I make a point of situating the philosophy in its personal context, emphasizing the people, places, and times involved. There is also an element of struggle inherent in living a life of philosophy, and I have tried to bring this out in my own case. The result, I hope, captures the texture of the day-to-day consciousness of a professional philosopher.

I should make it clear that this is not a full autobiography; it is an intellectual autobiography. I therefore mention only those experiences and relationships that bear directly upon my intellectual life. I make no mention of experiences and relationships, however important they may have been, that do not contribute toward a portrait of my philosophical existence. Mostly, this is a book about what has gone on in my mind. I want to stress this because it might be only too easy for my reader to interpret a lack of discussion of extra-philosophical events as a lack of the relevant kinds of events. One stereotype of the philosopher has him or her confined to a lonely room all day, ceaselessly contemplating—as if there were no love or sex in the life of a dedicated philosopher. I grant that the way I describe my life here might confirm this stereotype, since I make no mention of such worldly matters. But, to repeat, the absence of such discussion should not be taken to imply the absence of the relevant kinds of experience—far from it. I have simply carved out that portion of my life in which philosophy has occupied center stage. There is already enough turmoil and drama there to be mentioned.

The kind of philosophy dealt with in this book is often labeled "analytical philosophy." This is really too narrow a label, since what I choose to discuss is not generally a matter of taking a word or concept and analyzing it (whatever exactly that might be). The philosophical tradition of which I am a part begins with Plato and Aristotle, continues through Locke, Berkeley, and Hume, as well as Descartes, Leibniz, and Kant, and reaches Frege, Russell, Wittgenstein, and their recent philosophical descendants. This tradition emphasizes clarity, rigor, argument, theory, truth. It is not a tradition that aims primarily for inspiration or consolation or ideology. Nor is it particularly concerned with "philosophy of life," though parts of it are. This kind of philosophy is more like science than religion, more like mathematics than poetry—though it is neither science nor mathematics. So readers in search of a book about eastern philosophy, or Continental philosophy, or "postmodern" philosophy, will not be rewarded by this one. This is a book about philosophy as it is practiced today in the university philosophy departments of the world and which is now largely the "analytical" kind of philosophy.

The book is thus about contemporary academic philosophy. What I have aimed to do is present an extended glimpse of this world, so that the reader will come away with an impression of what the genuine article looks like. Some of my discussions may therefore seem tough going, and removed from the concerns of everyday life, but I have tried to make the ideas as clear as possible while doing justice to academic philosophy as it has actually been practiced over the last thirty years. What you get here is what there is. Not that this kind of philosophy is somehow only of interest to those who are paid to practice it; part of my point in this book is that these ideas should be of interest to everyone who is concerned with fundamental questions about reality. As I am at pains to show, I myself was drawn to philosophy—irresistibly so—simply by the intrinsic interest of the questions; in a way, it is accidental that I ended up teaching in a university philosophy department.

And let me not give too somber and heavy a picture of what follows. It is not all brow-creasing deep thought; philosophy has its funny side, and its all-too-human side. Philosophers, after all, are human beings too, surprising as that may sound.

Colin McGinn
New York
July 2001

THE MAKING OF A
PHILOSOPHER

First Stirrings

I WAS BORN IN 1950, FIVE YEARS AFTER THE END OF WORLD WAR II, in West Hartlepool, county Durham, a small mining town in the northeast of England. The hospital in which I was born was a converted workhouse, or homeless shelter as it would be known today. My mother was twenty years old, my father twenty-six, and I was their first child. Both my grandfathers—whose names were both Joseph, like my father's—were coal miners, as were all of my uncles except one, who was a carpenter and bricklayer. Life expectancy among miners was low, and both my grandfathers died young from work-related diseases. Everyone in my family was short and wiry. My paternal grandfather was known in the mine as "Joe the Agitator" because of his activities in fighting for improved working conditions; he eventually became secretary of the local miners' union, and read Karl

Marx and Rudyard Kipling in his spare time. He was a kindly, clipped man, not much given to conversation, devoted to his Woodbine smokes. I never remember a time when my tiny, shrill-voiced, constantly cussing grandmother had any teeth; she chewed meat with her gums. She said "thee" and "thou" (pronounced *thoo*) as part of ordinary speech, as in "thee knaas Jack Ridley" (meaning "you know Jack Ridley"). Of a blunt knife she would say "I could ride bare-arsed to London on this" and give out a throaty, high-pitched cackle. I have no recollection of my maternal grandfather, though his widow is still miraculously alive at ninety. My father left school at fourteen and went "down the pit," his first job being to pick stones out of the coal as it was shunted by on a massive belt contraption. But he quickly escaped this form of premature burial by going to night school and learning the building trade. He was sufficiently proficient at this to become general manager of a small building company while still in his twenties, and he made his career as manager of various branches of the building department of the co-op in different parts of England. He retired early and now has a second career as a painter, mainly of scenes from the mining towns in which he grew up. Some of his work is in the historical record of the art gallery that serves the area his paintings record. Both my brothers, Keith and Martin, are artists too, though I was never very strong in that department.

I have no recollection of my first three years in the northeast, and when I was three we moved to Gillingham, Kent, in the southeast of England. What a difference three hundred miles makes. Kent is known as the "garden of England," while county Durham was a place of enormous smoking slag heaps, cramped terraced streets, and chilly outside toilets. In Gillingham I enjoyed the woods and the fields, taking a special interest in wildlife—particularly lizards and butterflies—and grew to be the tallest McGinn on record (I am five feet six inches tall), —until my giant of a younger brother took over at a remarkable five feet nine. At age eleven I took the infamous Eleven Plus, a scholastic test to determine what type of school you would go to for the rest of your school years, and did not perform well enough to go to a grammar school. I was therefore sent to the local technical school, where I was expected to learn the skills necessary to become a tradesman or technician. However, after only eight years in Gillingham we moved again, this time to Blackpool in the northwest, and after a series of mishaps I was sent to the local secondary modern school—one step *down* from the technical school in the south.

Blackpool is a rough, tough, garish seaside town, windy and wet, frequented mainly by working-class people taking cheap trips. Its streets are lined with pubs, fish-and-chip shops, and amusement arcades. Cultural it is not.

And yet there abided an odd sense of privilege in the locals, even a kind of snobbery, since people did actually choose to pay good money to visit the place. The main activities of young men in the town were drinking and fighting, and trying to take clumsy advantage of visiting girls under the piers. The school I attended was loutish and philistine, mainly an exercise in crowd control, though frequently hilarious (the tubby headmistress was actually named Miss Bloomer—"Keks" to the boys, local dialect for "knickers"). On one occasion the PE teacher caned an entire year of boys—some ninety behinds—because someone had thrown potato crisps over the locker-room wall at the swimming pool and no one would reveal the identity of the culprit. I was caned three times in all, the other two times for no particular reason either (and it really stings too). It was not a school from the experience of which you were expected to amount to anything; most of the boys I knew there were in low-level jobs by the age of sixteen. Still, I always did pretty well in mathematics and English (but boy, was I bad at geography). I made a point of getting my homework over with as quickly as possible and spent most of my time on sports, playing drums in a rock band, and perfecting my pinball skills.

I did, however, perform well enough in my O-levels, taken at age sixteen, to be transferred to the local grammar school to study for my A-levels. Here I was spectacularly outshone by my classmates, who struck me as virtual geniuses, comparatively speaking. Some of these boys actually read books for pleasure! I was a big reader of children's books when I was young, especially the Dr. Doolittle stories, but since adolescence I had read almost nothing, just the odd horror story or piece of science fiction. Reading had lost its magic for me at around age twelve, when, coincidentally, the hormones kicked in. What I was good at, and enjoyed, was sports, especially gymnastics and pole-vaulting (for which I held the school record). I was also part of the in-group of "mods," who paid particular attention to their hairdos and clothes (backcombed hair sculptures, sharp suits, dancing shoes). At this time I had no thought of going to university, and the idea had never been mentioned in my house; it was not something a McGinn had ever done before. My teachers expected that I would become a PE teacher because of my sporting talents and moderate ability with book learning. My own thoughts turned rather to becoming a circus acrobat or professional percussionist. But one day at school we were asked whether we wanted to take a shot at going to university and I figured it might be worth a try. And anyway big changes were already under way in my mental development. My life started to shift to my head, at least in part. Up to now, developing physical coordination had been my chief concern, what with the sports and the drumming, but now my mind started to crave activity too. It was like a switch being turned on: The circuits began to hum.

I had fallen under the influence of a teacher, Mr. Marsh, who taught me Divinity A-level. I had already been much impressed by the intellectual adventures described in James Joyce's *Portrait of the Artist as a Young Man*, which was part of our prescribed reading for English A-level (my third A-level was Economics). But Mr. Marsh ignited in me an interest in studying and thinking, particularly about religion and theology. He was a strict teacher, but kindly—a devoted Christian with a passionate interest in his students. As I look back, he strikes me as a man who loved learning and scholarship (his favorite word was "scholar") but who didn't quite have the ability to make it as a university professor. He spoke of his university days as though they were a veritable heaven, his eyes burning with remembered enthusiasm. He taught us the Bible with great intensity, but not as a proselytizer—he had a genuine fascination for theological questions. He would occasionally mention philosophers as he was discussing some contentious point—as it might be, the plausibility of the virgin birth—and from him I first heard the name of Descartes.

Descartes was described as sitting in his oven on a cold winter's day doubting everything, even the entire external world and the existence of minds other than his own. All that was left was his own self as a thinking being. This was meant to demonstrate the futility of doubt and the importance of faith: If you doubted the events of the Bible, you would end up doubting everything. In the end, Mr. Marsh triumphantly argued, Descartes could believe only in his own existence as a solitary mind—that is where doubt would lead you! This was very strange—dramatically opposed to common sense—and yet there seemed to me to be a logic to Descartes's doubts, whatever their bearing on religion might be (in fact, Descartes's system relies centrally on God, but Mr. Marsh never mentioned this).

As a result of these philosophical intrusions I started to dip into some elementary philosophy books (if there can really be said to be such things). Naturally, I was very concerned with whether the existence of God could be rationally established, especially since at that time I would have counted myself a Christian believer: not that I had been brought up this way, but studying the Bible under the enthusiastic Mr. Marsh had led me to these beliefs. And once you believe in God, with all that this implies, you become curious about the intellectual foundations of the belief. Is it just a matter of blind faith or can God's existence be proven? And asking this question quickly leads to the whole issue of what a justification is anyway, as well as to questions about knowledge, certainty, free will, and the origin of the universe. God may or may not be a philosopher, but he is certainly responsible for a lot of philosophy.

And here is where my very first philosophical epiphany occurred. I was sitting in my cold, unheated bedroom in Blackpool, my drums in the corner, quietly reading a book about arguments for the existence of God (I forget now what book it was). I came across something called the "ontological argument," invented by Saint Anselm of Canterbury in the Middle Ages. I found the argument hard to follow but absolutely riveting (a lot of philosophy is like that). I kept reading the words over and over again, trying to absorb their meaning, as my feet grew colder. The sensation was of my mind being seized by abstract reason and carried willy-nilly by the power of logic. The ontological argument goes like this: God is by definition the most perfect being of which you can conceive. He combines all the perfections in one entity—absolutely good, perfectly wise, infinitely powerful. This is just what we mean by the word "God," and apparently we can mean this whether or not God actually exists. As Anselm put it, God is defined as the being "than whom no greater can be conceived." That is, *if* God exists, then by definition he is the sum of all perfections—just as, if a unicorn were to exist it would have a single horn. The question put by someone who doubts God's existence is whether there exists anything in reality answering to this definition. Yes, God *would* be the most perfect existing being *if* he existed: but does he? After all, I can define a word "Gad" to mean "the person who can jump bare-footed higher than thirty feet in the air with the greatest of ease," but that doesn't tell us that Gad really exists—and in fact there is no such person as Gad. The question of God's existence is analogous, it might be thought; we know the definition of the word "God," but what we don't know is whether there is anything in reality that answers to this definition. An agnostic who doubts God's existence surely knows perfectly well what the word "God" means—just as we all know what "unicorn" means. So at least it might be thought that atheism is a logically consistent position; it's not like claiming that triangles don't have three sides, which is false by definition. The question of God's existence is a question of fact, not a question of mere definition.

But, Anselm argues, this is wrong: Atheism is *not* a logically consistent position after all. Why? Because we are forgetting that God is defined as the most perfect conceivable being in *every* respect—and is it not better to exist than not to exist? If God does not exist, then he lacks the attribute of existence; but then, isn't he less perfect than a similar being who has this attribute? Take two beings who are alike in their perfections, except that one exists and the other doesn't exist: Doesn't the existent being have more perfections than the nonexistent being, since he at least exists? Not to exist is a kind of failure, a lack, but God is *defined* as the being who fails at nothing, who lacks no positive quality, who gets everything right, who has it all. Such a being *has* to exist or else he fails to have every positive quality. So the existence of God does follow from the definition of God, unlike with my case of Gad, the nonexistent high jumper. Once you know what the word "God" means you thereby know that God exists, since what we mean by God is just the most perfect being, and existence is one of the perfections. Existence is an attribute that augments or increases an entity's degree of perfection, so the most perfect conceivable being must have this attribute.

Consider the idea of the most powerful conceivable entity: Doesn't such an entity *have* to exist, for the simple reason that not existing is a drastic reduction in how powerful an entity is? To put the argument in the terse classical form in which I first encountered it: God is defined as the being than whom none greater can be conceived; but existence is an attribute that contributes to greatness; therefore God exists. God thus exists by virtue of the meaning of words, as a kind of conceptual necessity; so it is not logically coherent to doubt his existence, as if this could be a separate matter from what we mean by the word "God." The existence of God is logically necessary, a matter of pure definition, not a matter of contingent fact. The case of God is therefore quite unlike the case of the unicorn, whose definition does not imply its existence.

Now, this is a stunning piece of reasoning. It purports to establish by rigorous logical argument that the existence of God cannot be sensibly denied. No need to appeal to leaps of faith or speculations about how the world began or the occurrence of miracles: We get the existence of God for free, as a matter of pure reason. To someone like me, at age eighteen, struggling with the question of God's existence, this seemed like a bolt from the blue. God's existence turns out to be as solid as the fact that four is the next whole number after three. But, as I studied the argument, rereading it, trying to probe its workings (my feet getting colder all the time), I dimly felt that somehow the reasoning was too clever by half, that it made the question of God's existence *too* easy, that it rendered faith irrelevant. So, while I was impressed with the argument, and for a while obsessed with it, it left me with a disturbed feeling. A lot of philosophy is like that: gripping, momentous, but also worrying, naggingly so.

I think what really shook me up that day was a sense of the power of reason—of how logical thinking can produce big, shocking results. It is not that I still believe that the ontological argument is sound, though I don't think there is anything *obviously* wrong with it. But it is a fascinating argument, simple yet intricate, and I am not now at all surprised at the impact it had on my eighteen-year-old self. On that day I knew that I wanted to learn more of this philosophy business. Apart from anything else, the argument was just so damn *clever*. Imagine how Anselm must have felt on the day that he invented the ontological argument; he must have walked around Canterbury in a daze of excitement and awe for weeks. (There was, unfortunately, no Saint Anselm of Blackpool, whose shrine I might visit.) In fact the argument was largely accepted by the major philosophers who succeeded Anselm, so it counts as one of the most influential philosophical arguments in history. What also impressed me on that wintry day in Blackpool was the fact that my mind could be put in contact with the minds of great thinkers from the past, and taken away from the humdrum vulgarities of the seaside town in which I happened to live. That peculiarly transporting quality of philosophy has always stayed with me, and I feel it even now as I type these words (also in a none-too-glamorous seaside town: Mastic Beach, Long Island). Philosophy can lift you up and take you far away.

At around this time I started reading books by C.E.M. Joad, at Mr. Marsh's suggestion. Joad wrote accessible philosophy books for the general public and used to speak regularly on BBC radio in the 1950s. He was not himself an original philosopher but derived most of his ideas from

Bertrand Russell, about whom more later. (Russell was once asked to write a laudatory preface for a book of Joad's and testily replied "Modesty forbids.") Joad's passion was perception, and he was a devotee of the "argument from illusion." The question at issue is whether we really see physical objects out there in space or just subjective items in our own minds. Normally, we suppose that we see physical objects all the time, and touch them too: We see trains and boats and planes, and we touch doorknobs, cups, and bodies. What could be more obvious and commonsensical? But for centuries philosophers were convinced that this was just a vulgar error, a mere manner of speaking that concealed the truth about perception. What we really perceive are elements of our own minds—variously called sense-data, impressions, experiences, representations. Perception is not a faculty that unlocks the ways of the physical world outside of our minds; rather, it is confined to a purely inner array of mental items. The only reality we ever literally perceive is a virtual reality, a shadow world of fleeting sensations. Hence the philosophers' description of the ordinary view of perception as "naive realism": We no more really perceive physical objects than the sun really rises or the earth is really flat—these are just naive illusions. What we perceive is inside us, not outside, as we naively think.

Why would anyone reject common sense in that way?

Here is the standard argument, propounded with great force and clarity by Joad, following a long tradition. Consider illusions and hallucinations, such as seeing a straight stick look bent in water, or imagining pink rats while under the influence of LSD. The stick looks bent in just the way a really bent stick would look, and the pink rats look just the way real pink rats would look, if there were any. Illusion is precisely a mimicking of reality, which is why it can take you in. So there is no subjective difference between illusory perception and real or "veridical" perception: The world looks a certain way to you, and this looking can be illusory or veridical, depending upon whether the world is really the way it looks. From a subjective point of view, there is no distinction between Macbeth's hallucinated dagger and a real dagger—which is why illusions can be as scary as real things. But now, in the hallucinatory case, you don't see any real physical object—there *is* no object in the external world that answers to your experience. Yet you surely experience something, your mind is not a blank—things do look a certain way to you. So in this case you must be seeing something other than a physical object, and this something must be mental in nature. Call this something a "sense-datum": so you can be said to see sense-data of bent sticks, pink rats, and daggers, even when you really see none of these physical things . You perceive the subjective appearances of things, not things themselves, since they aren't there to be perceived. What you are aware of in a case of illusion or hallucination are nonphysical internal sense-data, not physical objects. But now, as we noted earlier, there is no subjective difference between the illusory case and the veridical case: Things look the same in both cases; your experience is the same; you can't tell the difference. Doesn't this mean that you must be aware of sense-data in the ordinary veridical case too? When you "see" a real dagger you *see* a sense-datum of a dagger; it is just that there is an actual dagger there that matches your directly experienced daggerlike sense-datum. So what you immediately perceive is *always* the sense-datum, not the real thing. Therefore you do not directly see physical objects at all but only their representatives in the shape of mental sense-data. It is like trying to meet the head of state and only getting as far as her emissaries. Your direct awareness stops at the level of sense-data and does not reach out and catch hold of actual physical objects. Maybe we can say that you *indirectly* perceive physical objects, as when you see only reflections in a mirror or somebody's footprints in the snow. But they are not what is immediately before your mind when you have a visual experience. What is immediately before your mind is your mind itself—its current sensory contents.

This is an alarming result. It restricts your awareness to your own subjective self, cutting you off from the world

of physical objects with which you naively thought yourself to be in contact. Consciousness acquaints you only
with its own contents, a play of images upon a mental
screen. When I first encountered this argument I would
stare at the furniture around me and try to force my mind
to become aware of it, to penetrate the veil of sense-data;
but I had the stifling feeling that I was only gazing harder
at what was inside me—my own subjective world, not the
common public world I had believed in up until then.
There was at best a correspondence between the subjective world I was experiencing and the physical world
beyond, but there was no way I could step out of my subjective world to check that the correspondence really
held—since I had no direct access to the physical objects
that supposedly corresponded to my sense-data. In a way
it was like discovering myself to be blind: I couldn't see
physical objects! Nor could I touch them, taste them, or
smell them. My world had shrunk down to my own conscious self. I was, I suppose, as self-absorbed as many
other adolescents, but this was too much. I had lost the
world, or rather I had never had it to start with.

And from there it only got worse. If we never perceive
physical objects, how do we know they are really there?
Not by perception, certainly. Normally, you think you can
tell what is in a room by going to have a look. But not so:
All you determine that way is what sense-data you have

after having the sense-data of walking to the room; the objects remain maddeningly out of reach. It is just a short step from this to full-blown industrial-strength Cartesian skepticism: All you really know to exist are your own subjective states and not the objects supposedly out there in the external physical world. Indeed, what right do you have to believe that there even *is* an external physical world? Might it not be the case that your sense-data don't correspond to anything external at all? Might not everything be a dream? Descartes asked us to consider an evil demon who produced sense-data in our minds but made sure that nothing ever corresponded to them, so that the entire course of our sensory experience is one long delusion. How can we rule out this possibility? Not merely by appeal to our sense-data, since they would be the same, whether caused by physical objects or by the evil demon. How do you even know that you have a body? How indeed do you know that you didn't come into existence five seconds ago equipped with an extensive range of pseudo-memories? Knowledge seems to shrink down to the inner states of a momentary self.

The updated version of Descartes's evil demon is the "brain in a vat" scenario. Here we suppose that our brains have been removed from our heads by alien scientists with highly advanced brain technology. The scientists put our brains in nutrient vats in nice little cubicles, each with our name on. Then they hook the brains up to a machine that sends electrical inputs into our sensory nerves, which result in sense-data: We experience familiar things but always by means of these electrical stimulations. Thus it feels as if you are in a bar in New York talking to your friends, but actually you are stuck in a vat somewhere in Cleveland hallucinating all this. What the scientists are doing is producing a mere simulation of the ordinary physical world—a virtual world of pure sense-data. They also cleverly monitor our decisions to move our bodies (which have been neatly disposed of) and give us sensory inputs that match what we decide to do. If I choose to go to the fridge for a beer, they gave me the feeling of my limbs moving and a kitchen coming into view and a fridge door opening, so that I cannot tell what my real situation is. I experience the bottle in my hand and then the beer going down my throat, but actually there is no beer, no hand, and no throat—just a brain with lots of wires sticking into it and a giant computer humming in the corner. Now the horrible question is: How do I know that I am not *now* a brain in a vat? If I were, things would not seem any different to me, so how things seem cannot rule out the vat hypothesis. I may at this very moment just be an immobile brain being fed the illusion that I am a walking, talking person in contact with real physical objects. Maybe I have never met a physical object in my life, including the physical objects we call people. Maybe solipsism is true and I am completely alone with my sense-data. Maybe the entire universe is just a shimmering wall of illusion, with my consciousness as the only reality. When I go, so does everything else, since there *is* no anything else.

I was pleased to see recently that this nightmarish vision had been converted into the basis for a sci-fi movie, *The Matrix*. In this film immobilized human beings are stacked in cubicles miles high and hooked up to computers that generate in their brains a simulation of the real world. The machines have done this to us in order to derive nutrients from human bodies; we have become computer food, factory-farmed. We accept this form of parasitism because we don't know that it is so, since everything *seems* quite normal to us. Only a small band of rebels know the true situation, and they are intent on liberating humanity. The entire premise of the movie is that if this were our predicament we would not know it: The Matrix would have us fooled. I would guess that the makers of the film had taken a philosophy course or two at college and been haunted ever since by the possibilities dreamed up by Cartesian skepticism. And the whole contemporary movement toward virtual-reality machines depends upon the possibility of simulating the real world, thus disconnecting how things seem and how they really are. After all, we experience what we do because of the signals that enter our sensory nerves and activate parts of our brain; if we can reproduce those signals without the aid of actual physical objects, then we can simulate experience of objects without bothering with reality at all. So far as experience is concerned, reality can be dispensed with, at least as a matter of principle.

Again, what struck me at age eighteen about these arguments was the power of reason to overturn assumptions I had taken for granted. It is not that I would now subscribe to these arguments in their entirety, but they are certainly logical enough—and they shatter a central part of our commonsense view of our place in the world. Our commonsense beliefs are not as rationally impregnable as we fondly supposed before we inquired into their foundations. The argument from illusion is like a tidal wave of reasoning that washes you up on a strange, alien beach, leaving you exhilarated but disturbed. And notice how that argument combines with the ontological argument: It turns out that reason can prove logically that God exists, but it also leads to the result that we cannot know that the ordinary world of tables and chairs exists. We thought that the existence of God was rationally questionable but that the external world was solid as a rock, yet reason tells us that we can be surer of God's existence than of our neighbor's existence. Divinities are more certain than tables and chairs! I can know for certain that I exist and that God exists, but every-thing else is doubtful. Who knew? What launched me on the road to becoming a philosopher was wanting to know whether this is really so; the arguments seemed com-pelling, but were they really as compelling as they seemed? I wanted to understand the arguments better, so that I could decide what to believe. What I am emphasizing now is the impact these arguments had on my receptive teenage mind—how they got me *thinking*. I began to real-ize that even the most familiar belief might be mistaken, a mere prejudice—that everything had to be open to rational scrutiny. If this was true for belief in the external world, how much more was it true for the kinds of social and polit-ical beliefs that were taken as gospel? Since all this was taking place in the late sixties, a time of revision and upheaval, it only added to my sense that everything was up for question (though it must be said that at this time I was not politically engaged).

I think that I was also attracted to Freud at this time because the power of reason could establish the unobvious, though I cannot recall now exactly what prompted me to read his works. Freud, as everyone knows, posited an unconscious mind operating behind the scenes and being responsible for some of the oddities of conscious life. Dreams, for example, have a certain content which we can often recall upon waking, but they often seem meaningless and baffling, like a code we cannot decipher. Freud offered to make sense of these hieroglyphics of sleep by postulating a hidden mental reality with its own logic and agenda—as he did with neurotic symptoms, slips of the tongue, jokes, and so on. By analyzing these phenomena and applying suitable theoretical constructs, Freud makes the meaningless intelligible, revealing the surface phenomenon as just the outward sign of a coherent hidden reality. Freud applied reason to make the irrational seem reasonable. In reading Freud's works at eighteen I experienced the pleasure involved in the activity of *making sense.*

Freud was not a philosopher, strictly speaking, but he shared with philosophers the drive to press reason as far as it will go, a readiness to question, and the desire to delve beneath the appearances. It was Freud who first got me interested in the workings of the mind, which became my central interest in later years. And it was Freud who led me to decide to study psychology at university. Despite my growing interest in philosophy during my last year at school, I did not apply to study philosophy at university, for the simple reason that I had no idea that it could lead to an actual job. At that time I had no idea that I would ever become an academic—I am not sure that I even knew what an academic was—and so I needed to take a degree in something that might lead to paid employment. Psychology offered that prospect: I thought of becoming an educational psychologist or a clinical psychologist, and my reading of Freud made me think that psychology would offer me the intellectual satisfaction I sought. I knew that I could always study philosophy as a subsidiary subject, and that seemed enough. I was, after all, to be the first member of the McGinn family ever to apply to university, and it would have seemed hopelessly impractical to apply to study philosophy, of all things. I was pushing the envelope to apply for psychology instead of economics, which was deemed a far more practical option by my family and by my teachers. (My father had expected me to become a quantity surveyor in the building trade because of my proficiency in English and mathematics, but that idea went by the wayside.)

Yet I had to get to university, and I was not doing terribly well at school. My weakest subject was economics, in which I was not far from the bottom of the class. I then hit upon a novel idea: Instead of just listening to the teachers in class I could study a textbook by myself at home. Mr. Marsh had given us some tips on how to study—make notes, reread, rehearse what you had learned in your own words—and I applied myself to an economics textbook with these elementary pointers in mind. At the time I had an early-morning paper route (so that I could pay for my secondhand drum set) and I would study economics in the hour between finishing the paper route and starting school.

I did this for a few weeks, and in the next economics examination I came third in the class. It worked! I did the same with Divinity and English, in effect teaching myself, putting in the hours, and by the time the A-level examinations came around I was able to obtain an A and two Bs (the A in economics). Nothing spectacular there, I admit, but considering where I had started from only several months earlier, not too shabby. And those grades got me admitted into Manchester University to study psychology. I was the first McGinn to go to university, and I was glad to feel that I would be able to study subjects that interested me for the next three years. I had gone from underachieving jock-mod to pocket-sized intellectual in less than a year, and philosophy had to take a lot of the blame.

Meanwhile, before we get to university in my story, picture me sitting on a bench staring at a British mailbox on a blustery spring day in Blackpool. I had just been reading about the question of substance and qualities, and was suitably transfixed. Is an object just the sum of its qualities or does it have an existence that in some way goes beyond its qualities? The mailbox had a variety of qualities—it was red, cylindrical, metal, etc.—but it seemed to be more than just the collection of these; it was a *thing*, a "substance," that *had* these qualities. But what was this substance that had those qualities? Did it lie behind them in some way, supporting them like the foundation of a house? If so, what was this underlying thing like—what qualities did it have? If it had some qualities, wouldn't there be the same problem again, since it would also have to be distinct from *these* qualities? But if it had no qualities, what kind of thing could it be? How could there be something that had no qualities? So maybe we should say that there is nothing more to the mailbox than the qualities it manifests. And yet how can an object be just a set of abstract qualities? Isn't it more solid and concrete than that? I was confused, and no amount of staring at the mailbox was helping me. I kept repeating these words in my mind (and I remember it so well, over thirty years later): Is the object something over and above its qualities or not? I had a vague mental image of a gray amorphous something that constituted the underlying mailbox, to which its various manifest qualities mysteriously were attached. But it seemed elusive and peculiar, a philosophical invention or monster. Yet as soon as I replaced this fuzzy image with the qualities by themselves, trying to think of the mailbox as just a "bundle of qualities," the object itself seemed to disappear. I was in the grip of a characteristically philosophical problem in which the simplest question produces conceptual bafflement. I didn't even know what an object was! Bafflement of this kind stayed with me, as a kind of ongoing perturbation of the mind, even as I applied myself to more practical concerns.

CHAPTER TWO

From Psychology to Philosophy

I LEFT WINDY BLACKPOOL FOR RAINY MANCHESTER IN THE autumn of 1968, with all the usual anxieties of a provincial boy going up to university for the first time. I was installed in the house of a Mrs. Readyoff, my landlady, with two other boys, sharing a bedroom with an electronics student from Wigan who had an accent even thicker than mine. Mrs. Readyoff would cook us huge, stodgy meals to be served promptly at five P.M., and we would spend the evenings chatting and studying (no TV in the house). I had enrolled for the BA in psychology, with a subsidiary introductory philosophy course. There was also a compulsory language requirement, which proved to be the bane of my life. Everyone had to take a language course and pass the examination at the end of the year if they were to continue at university. Most people took

French as the easy option, having already studied French at O-level. But my secondary modern school in Blackpool did not offer O-level French, so I had no background in it; I had O-level woodworking instead. I therefore enrolled for a language that was to be taught from scratch, Italian, thinking that it would be suitable for someone like me with little background in languages. I could not have been more wrong. As I discovered once the term started, the French course was designed for language dunces in order to fulfill the language requirement, while the Italian course was for language whizzes wanting to add another language to their quiver, and most of the other students in the class were taking French or Spanish as their main degree course. We were expected to master all of Italian grammar and vocabulary in the first ten-week term, meeting twice a week for an hour; and in the second term we had to study a nineteenth-century Italian novel (*Il Cappello del Prete*) for which there existed no English translation. It was, I have to say, incredibly difficult. I failed the examination at the end of the first term abysmally, which didn't surprise me a bit, despite some remedial help from the generous teacher of the course. I was thus faced with the prospect of being ejected from university if I failed the examination at the end of the year. I accordingly spent most of my time studying Italian, memorizing grammar and vocabulary—memorization not being my strong suit at the best of times. During class the teacher would always turn to me and ask, *"Che ora è, Signor McGinn?"* ("What is the time, Mr. McGinn?") and I would fumblingly reply in Italian, to the merriment of all present. Self-esteem? Not in those days.

There were also compulsory weekly lab reports for psychology, so that in the end I spent almost no time on philosophy, which I was desperate to study. Fortunately, I passed the Italian exam at the end of the year, by dint of immense effort (it included such choice items as translating a difficult passage from James Joyce's *Ulysses* into Italian). I still think that mastering Italian during that year was the most difficult intellectual task of my entire life, and I have never read an Italian menu without a silent shudder. I have now forgotten most of it, though I still know the rules of pronunciation perfectly; at least I know precisely how to pronounce *cognoscenti*, a useful word in my line of business.

My studies improved during my second year at Manchester, relatively speaking. I ended up living alone in a tiny, depressing bed-sitting room in a rough part of the city, cycling to classes in the rain on a bike that went wrong all the time, but at least I had more freedom to pursue my interests. Since I had no money there wasn't much to do except read and talk (does poverty encourage literacy?), and I made friends with some people of similar inclinations. We used to sit up till four in the morning, discussing the methodology of psychology, drinking cups of instant coffee with our coats on (proper heating was far too expensive). By now I was taking a course in the philosophy of science and reading much more widely in the philosophy and psychology. But there was a problem: I was too interested in philosophy and not enough in psychology. Psychology began to feel like a bit of a grind, while philosophy was thrilling and gripping to me. But I was reluctant to switch my degree to philosophy, even if I could, because I felt there was no possibility of a career there; so I had to stick with psychology. I would have liked to combine the two subjects in a joint degree, but Manchester happened not to allow that particular combination of subjects in its joint degrees. Meanwhile I was doing quite well enough in psychology and not finding it wholly unsatisfying. I particularly enjoyed devising experiments and carrying them out, developing a special interest in visual search—the process by which you pick out a particular target stimulus against a background of other stimuli, as when you search for a familiar face in a crowd. But my interests kept returning to foundational issues, and I became increasingly frustrated, especially since my psychology teachers took a dim view of philosophy as "unempirical." I think they saw me as a traitor to scientific psychology and disapproved of my philosophical

predilections. Anyway, I resolved to complete my psychology BA and take it from there; so I persisted with the
IQ tests, the brain physiology, the reaction-time experiments, the industrial psychology course.

For some reason the head of the psychology department, Professor John Cohen, took a shine to me and
would always stop to chat with me when our paths
crossed. I was nineteen and he was over sixty at the time,
a man of extensive learning and broad intellectual sympa-
thies, short and compact like me. I think he sensed my
intellectual enthusiasm, and we shared a sense of humor.
He was disappointed when I eventually defected to philosophy, but we kept up our convivial contact until his
death in 1986. I felt a great affection for him and immense
gratitude for his encouragement and attention (I was, I
think, the only student he ever invited to his house). He
helped give me confidence in myself. It is hard to exag-
gerate the importance of this type of contact between students and teachers; for you to believe in yourself, someone else you respect has to believe in you first. As a
teacher now, I try to keep my eye open for students who
could benefit from this kind of attention and encouragement, while avoiding favoritism and excessive "mentor-
ing." A wink and a few words can be enough. (I always
called John Cohen "Prof," even long after he ceased to be
my teacher. My letters would begin "Dear Prof"; I could

never bring myself to address him as John, and my use of "Prof" was a kind of joke between us.) He was the first serious intellectual I got to know well.

Aside from this real presence there was another remote presence in my mental life: Bertrand Russell. I find it hard to capture just how powerful the Russell effect was on me, and I suspect on many others. Nowadays I have a much more qualified view of him, but to me back then he represented an ideal of intellectual and moral authority. Reading his three-volume autobiography was one of the formative experiences of my young life. It begins, memorably enough, "Three passions, simple but overwhelmingly strong, have governed my life: The longing for love, the search for knowledge, and unbearable pity for the suffering of mankind. These passions, like great winds, have blown me hither and thither, in a wayward course, over a deep ocean of anguish, reaching to the very verge of despair." And so it goes on, steeped in Russell's particular blend of passion and precision—or, as one might less favorably put it, madness and mathematics. The features of Russell's writings that impressed me the most were his enormous intellectual ability, his unswerving commitment to reason, and his marvelously pellucid prose style. He made the life of the mind seem like a heroic adventure, not a monkish confinement to dusty libraries. The ten years he took to write the massive *Prin-*

cipia Mathematica (with A. N. Whitehead), working ten mathematical hours a day, struggling with his famous paradox of the class of classes that are not members of themselves, seemed like an epic expedition across uncharted terrain, icy but beautiful. There was nothing nerdy about Bertie. I also admired his wit and honesty; even his suffering (which was plentiful) seemed magnificent. More than anything else, it was reading him that persuaded me that I wanted to become a full-time, card-carrying philosopher. Moreover, Russell was a famous public figure, a winner of the Nobel Prize for literature, an early opponent of nuclear weapons, a regular speaker on radio and television, a scourge to the political establishment. I decided I wanted to meet him and planned to go down to Wales where he lived and request an audience during the Easter vacation. But he died, at age ninety-six, just before I could carry out this plan, so I never set eyes on him. I felt real grief when I heard of his death, and a sense of personal loss. In his writings I felt he was speaking directly to me. I even wrote to him once asking for advice on what to read; I received back a letter dictated to his secretary, but, alas, unsigned by him, advising me to read in many fields—I treasured it.

There was another side to my infatuation with Russell (whom I referred to as "BR," as though his name was too sacred to be mentioned), which has only recently

occurred to me. I am a child of the sixties, so I was reaching maturity just as rock 'n' roll was achieving cultural
domination. Like most other teenagers at this time, I was
totally taken with rock 'n' roll and the people who created
it—Elvis Presley, the Beatles, the Rolling Stones, the Who,
and many others. As I said in chapter 1, I played drums in
a band (with my brother Keith) from about fifteen to eighteen, doing gigs in local church halls and the like. Yes, I
confess, I have signed my autograph on girls' arms. Many
of the stars of rock 'n' roll were provincial working-class
boys like me, and it was a lifestyle which had a tremen-
dous appeal to me—and something I could easily identify
with (more easily than being a philosopher). If my band
had been more ambitious and successful, I might really
have had a career as a rock drummer. But the mentality of
the rock 'n' roll life stayed with me long after I stopped
performing (I still have a pristine set of Premier drums in
my study), and my infatuation with Russell obscurely
connects with that mentality. Russell had a kind of rock-
star philosopher's life: He made philosophy do for him
what music did for, say, John Lennon. They were both
romantic public figures, heroes of youth, vocal critics of
the establishment, masters of the word. And just as the
life of John Lennon had a great appeal to me as a sixties
boy, so did Bertrand Russell's—though with obvious differences. Lennon and Russell had eloquence, talent,

courage, and style. I still see philosophy that way: not as the pursuit of dried-out bookworms with no life, but as a life of creativity, commitment, and independence of spirit. There is something *radical* about philosophy—audacious, potent, fundamental.

What of my religious beliefs? It was Russell who extinguished the last remnants of religiosity from my soul. I had already been thinking of human beings in a more scientific spirit than I had at school, partly as a result of my psychological studies—the structure of the brain and its indispensable role in governing mental life, the influences of heredity and environment on character, the continuity between animals and humans. This naturalistic view of man was in contrast with the religious image of an immaterial soul that can survive bodily death and is not subject to the laws of nature. I also became increasingly disturbed about the possibility of free will, which is presupposed by the entire Christian conception of praise and blame, reward and punishment, heaven and hell. Like many another before me, I could not for the life of me see how to reconcile free will with determinism: If all human action is determined by the laws of nature, so that every decision has its antecedent cause, how is it possible for the human will to be free? If the brain determines the mind, and the brain's actions are fixed by antecedent causes and laws, how could there be such a thing as free choice? And if character is determined by a mixture of heredity and environment, how can we hold someone responsible for his or her character, whether virtuous or vicious? This was not a purely intellectual problem to me, because of its crucial relevance to the Christian idea of moral responsibility, and I obsessed about it day and night. Russell, too, stressed the role of universal causality in determining human actions, and he openly defended atheism as the only rational position. His famous essay "Why I am not a Christian" examines all the classic arguments for the existence of God and ends by asserting that there is no more reason to believe in the Christian God than in the old, discredited Greek gods. The ontological argument, which had so entranced me a couple of years earlier, was dismissed by Russell as depending upon the mistaken logical idea that existence is an attribute like other attributes (a diagnosis I do not now quite agree with but found convincing at the time). He also dismissed the old argument that the universe needs a first cause and God must be invoked to play this role, for we can also ask what caused God, and this launches us upon a vicious regress of first causes. But these arguments merely cemented a process that was already under way. Religion simply lost its grip on me after a couple of years of fairly fervent belief. I shed it like an old skin; it slid off me quite naturally and painlessly. And of course, the strongest nonbelievers are always the ones who most strongly believed to begin with. I still admire many of the teachings of Jesus Christ, and find his life exemplary of some important moral truths, but I long ago rejected the supernatural baggage that accompanies Christian belief. To this day I have no belief whatever in the supernatural of any stripe—ghosts, psychics, telepathy, miracles, faith healing, and so on.

Russell also introduced me to what is now known, somewhat misleadingly, as "analytical philosophy." He had originally been a mathematician and had a deep understanding of modern science. His most important work was in mathematical logic, of which he was one of the main architects. He wanted to bring this kind of rigor to traditional philosophical questions, making philosophy more like mathematics and science and less a matter of airy speculation. Philosophy could thus become a hard-headed discipline, in which real progress could be made, instead of being a swamp of obscurity and pointless wrangling. This idea appealed to me (though now I would be far more hesitant about such a scientistic picture of philosophy). Unfortunately, some of Russell's work in this mode was very hard to understand because of its technical apparatus; *Principia Mathematica* consists of three volumes of virtually uninterrupted symbolism. In any case, it certainly looked rigorous, even if I couldn't follow most of it—nothing to be intellectually ashamed of. There had always been two sides to my mind, corresponding to my twin abilities in mathematics and English, and Russell's mathematical philosophy appealed to the mathematical side. The literary side of my mind also found a delight in Russell's elegant and powerful prose style (though his sporadic efforts at fiction were not a success, reading like fictionalized lectures more than anything else). I tried to adopt his mental outlook of passionate skepticism and analytical precision, and I even started smoking a pipe like him (he in his nineties and me a mere twenty). I well remember the day I finally tracked down the particular tobacco he favored, Golden Mixture, and started smoking it, expecting that it would make me as clever as he (it didn't and I gave up the bad habit of pipe smoking a year or two later). All in all, Russell has a lot to answer for in influencing a young lad from the northern provinces so thoroughly—and all done through the power of the written word. He probably did more to shape me than any other influence in my life.

I had another notable teacher at this time, Dr. Wolfe Mays, from the philosophy department. I first encountered him during a philosophy of science course I was taking during my second year at university. Dr. Mays was a conspicuously short, elfin man with one of the largest noses I have ever seen on a human being, the nostrils being a good inch and a half long. He always wore a dark gray suit, sported a halfhearted comb-over, and had an accent that seemed to be a mixture of cockney and Cambridge ("Can OY heve ve followin' number" he would intone into the phone). He was fond of boasting about his connection to the famous Swiss psychologist Jean Piaget, and once dismissed John Cohen's visit to Geneva to see Piaget as just "a gnat frew a window," compared to his own intimate involvement with the great man. We students really had a difficult time controlling our giggles during his lectures sometimes—as when someone's wooden clog accidentally fell to the floor and he looked up startled from his notes and querulously shouted "Come in!" But he was a man of considerable erudition and had a contagious enthusiasm for ideas, and I took to him immediately (remember, I am a short, beaky character also). He cared about his students and was very kind to me in a sort of intentionally offhand way (he always called me simply "McGinn"). I used to like visiting him in his office for a quick philosophical interchange, and never left without a Maysian gem of some sort ("OY'm workin' against toyme," he once said to me with great drama as he was preparing a reply to some other philosopher).

Aside from philosophy of science, Dr. Mays was a devotee of existentialism and phenomenology. In my third year at Manchester I took a course from him on Jean-Paul Sartre's *Being and Nothingness*, mainly because Dr. Mays was teaching it. There were about five students in the class and we proceeded by reading the book aloud and then going into exegesis. Now, this book is a tough read, being full of sentences like, "The Being by which Nothingness arrives in the world must nihilate Nothingness in its Being, and even so it still runs the risk of establishing Nothingness as a transcendent in the very heart of immanence unless it nihilates Nothingness in its being in connection with its own being. The Being by which Nothingness arrives in the world is a being such that in its Being the Nothingness of its Being is in question." Got it? But Dr. Mays would cut through the verbiage and give us a down-to-earth para-phrase of what we had just uncomprehendingly read. His characteristic mode of questioning would be to ask us if we knew the difference between X and Y, as in, "Does anyone know ve difference between ve For-itself and ve In-itself?" The alert student might reply, "The For-itself is human consciousness and the In-itself is the world of inanimate things," and Dr. Mays would explosively respond, "Vis is the ve difference!" By this means I managed to get through most of the text of *Being and Nothingness* and came away with a good understanding of its contents. And I found myself even starting to think in the Sartrean mode. This was very confusing, because a cornerstone of Sartre's philosophy is that man is radically free, and I had only lately concluded that free will is an illusion. The truth is that I felt compelled by both points of view and had no clear idea of how to reconcile them; I contented myself with the compromise position that we are free in one sense but not in another. This kind of conflict is quite common in philosophy, but it was especially acute in the present case; Sartre's analysis of human consciousness as a spontaneous emptiness seemed right, and yet there were all those arguments about determinism and the brain to contend with. It was all very confusing.

The basic Sartrean thesis is that human consciousness consists in its awareness of things outside itself: The being of the For-itself consists in its relation to the In-itself, to use the Sartrean jargon. To introduce another technical term that I will be needing later, consciousness is essentially constituted by its "intentionality." The concept of intentionality has nothing particularly to do with intending something; it is a technical term for the capacity of the mind to be aware *of* various things. If I see a glass of water or think about a glass of water or desire to drink a glass of water, then my mind is exhibiting intentionality in relation to glasses of water: The glass of water is the "intentional object" of my various mental states, the thing my mind is "directed toward." Sartre's thesis is that every act of consciousness involves intentionality, so that its essence is to be directed to what it is not—"it is what it is not," as he provocatively puts it. But then, he contends, the essence of consciousness must be nothingness, since there is nothing left over once you subtract the intentional objects from consciousness. Consciousness is pure directedness to objects, with no inner nature of its own, no intrinsic essence. So this is what we fundamentally are as centers of consciousness—nothingness. And it is this nothingness that lies at the root of our freedom: Because consciousness is nothingness we have no nature; we are just the free play of consciousness upon the world—consciousness on a stick, as it were. There is always a distance between consciousness and what it is conscious of, since consciousness does not entirely collapse into its objects, and in this distance lies our freedom. Hence the existentialist slogan, "Existence precedes essence." We are first of all a pure freedom, and only as a result of our choices do we acquire a nature. Character is chosen.

Now, there is a lot more I could say about this philosophy, much of it critical, but I think I have said enough to explain the appeal of the existentialist philosophy to me as a young man. It is a liberationist philosophy in which nothing can hold human beings back from becoming what they choose to be. Clearly, a young person in the process of constructing a life—an "identity"—will find much of resonance in the Sartrean picture, however problematic some of the central claims seem to be. The idea of Bad Faith as a denial of one's essential freedom, as a lapse into an identity imposed from the outside, has a powerful appeal for anyone struggling to forge an authentic and original self. Sartre's famous French waiter, who absorbs his role so thoroughly into himself that he acts like a machine or a caricature, has his counterpart in the young person who is expected to develop in a certain way, given his or her background, but feels the urge to cast off expectations and forge a new identity. Sartre's ponderous and paradoxical formulations thus had an obvious personal meaning for me, as I wondered what to do with my life (maybe I should make good use of all that radical freedom Sartre attributed to me). It was time to transcend circumstances and remake myself in the image I desired according to the values I chose. Whatever essence I was to have would be of my own choosing.

But I also derived something more abstract from studying Sartre; I deepened my interest in the philosophy of mind, particularly consciousness and intentionality, which was to flower in later years. Instead of studying the mind empirically, by means of experiments and surveys, one could study it analytically—by thinking about it systematically. Sartre had a system in which basic concepts led on to further conclusions, and he had an overall theory of the mind founded on phenomenological analysis (that is, analysis of consciousness as it presents itself to the subject of consciousness). This idea of a rigorous analytical system had gripped me in my studying of Russell on philosophy of mathematics, and Sartre, for all his differences from Russell, also had an analytical system of his own for the mind. The idea of axioms and deductions, systematically laid out, was immensely appealing, and I took great pleasure in seeing what followed logically from what. Perhaps that is one of the primary pleasures of philosophy: arriving at an idea and figuring out what its logical consequences are. This is much like the pleasure of archaeological excavation—you dig deeper and deeper into the conceptual soil, seeking intellectual treasure. The power of one idea to lead to another is a never-ending source of fascination for me. This is why the ontological argument for the existence of God is so quintessentially philosophical; from a premise about the definition of the concept of God it seeks to derive the conclusion that God actually exists, and the question is whether that conclusion really follows. It is always a damning moment when one philosopher says to another, "Hang on, that doesn't really follow." Since philosophy is largely about the construction of arguments, the philosopher needs to be finely tuned to what follows from what. "Non sequitur" is the ultimate put-down (P. F. Strawson once stingingly said of an argument of Kant's that it was "a non sequitur of numbing grossness").

I graduated with a First Class degree in psychology

from Manchester University in 1971. By then I had decided that I wanted to change from psychology to philosophy as a postgraduate. I had heard that Oxford was the best place to study philosophy in England, and they offered a two-year postgraduate degree, the B.Phil, that required taking courses in the core areas of philosophy. Since I hadn't covered most of the core areas of philosophy this sounded ideal, and I fondly imagined that the course must have been designed for people like me with little undergraduate training in philosophy. I could not have been more wrong (again). In fact, the B.Phil was designed for crack philosophy students who want to study philosophy in more depth at the postgraduate level, with a view to becoming professional academic philosophers. Blithely I applied to Oxford and found myself accepted by Balliol College, but the central university committee responsible for graduate philosophy admissions turned me down, to my great disappointment. Had I but known, this was not very surprising, since the B.Phil is a demanding degree which only the most philosophically able students can complete satisfactorily. (Not to keep the reader in suspense, I ended up many years later as an examiner of the B.Phil degree.) So my hopes of studying philosophy full-time were temporarily dashed. I decided to stay in Manchester for a year to take an MA in psychology, and then try again for Oxford the next year. This was not as frustrating as it sounds, because I had hit upon a topic at the border between psychology and philosophy that I wanted to work on more deeply: innate ideas.

Here is where the third greatest influence in my undergraduate life comes in—Noam Chomsky. Chomsky was a young man, in his thirties at the time, who was in the process of revolutionizing linguistics, and for good measure psychology too. It was de rigueur to be seen reading his *Syntactic Structures* and *Aspects of the Theory of Syntax*, both quite technical works. The main significance of Chomsky for me at this time (I shall be returning to him later) was his opposition to behaviorism. In the late sixties behaviorism was still the dominant ideology in psychology; it says that psychology is concerned with investigating external behavior, not the mind itself, which was declared hopelessly "private" and hence not open to scientific study. The basic model of behavior was held to be the conditioned reflex, which had been extensively investigated in rats and pigeons. Human learning, in particular, was supposed to consist in the inculcating of conditioned reflexes, in which "responses" come to be associated with "stimuli." Children were held to learn what they do by acting randomly and then having certain "behaviors" rewarded ("reinforced") and other behaviors punished ("inhibited"), thus selecting the former and eliminating the latter.

Behaviorist theory was roundly demolished by Chomsky, particularly for the case of the learning of the language. Chomsky convincingly argued that children could not learn language this way, indeed that they did not *learn* language at all; rather, they come to a specific human language already equipped at birth with an implicit grasp of the grammatical rules of human language. These rules are universal throughout languages and hidden deep beneath the overt structure of language. The job of psychology is to uncover the rules and mechanisms that underlie this innate linguistic competence; in other words, to work out the structure of an internal cognitive system. Astronomers try to figure out the makeup of the interior of the sun by observing its behavior and constructing explanatory theories; psycholinguists, Chomsky said, should try to figure out the contents of the innate cognitive system that constitutes linguistic competence and explain how it interacts with a specific language to produce ordinary mastery of English, or French, or whatever. One of Chomsky's key arguments was that the child has only limited and fragmentary data available from which to project a mastery of language—far too little to ground the rich grammatical competence that underlies speech—so that we have to assume that he or she brings to the task a rich, articulated system of linguistic competence in advance. In other words, we are born knowing the rules of grammar, and these rules are stored somewhere in our mind/brain. On the basis of these rules we can understand a potential infinity of new sentences that have never been uttered before; our linguistic mastery is thus generative and systematic. The child is not a "blank tablet" awaiting the imprint of the environment, but brings a sophisticated set of cognitive structures to the task of learning language. Kids start off a lot brainier than we thought, a lot more clued in.

This conception is the groundwork of the modern discipline of cognitive science, which I shall discuss later, but at the time it was shocking and revolutionary. Chomsky explicitly related his new ideas to the older tradition of rationalism, which flourished in the seventeenth century in the writings of Descartes, Leibniz, and others, so that he was advocating a return to insights that had been lost when behaviorism achieved dominance at the beginning of the twentieth century. He wanted a return to a properly scientific mentalism, in which the mind itself was the object of psychological study. This served to connect psychology and linguistics with the philosophical tradition and thus reestablished contact between these two worlds.

All this seemed very good to me (and still does). I wanted to write my MA thesis around these ideas, applying them to the question of mathematical knowledge: Must we also assume an innate grasp of mathematical principles in order to explain the acquisition of mathematical knowledge, as the seventeenth-century rationalists had maintained, or could we make do with empiricist principles of learning in which everything comes from experience? I argued that a position like Chomsky's was appropriate for mathematics too; we have innate mathematical ideas, a genetically fixed mathematical competence. I completed this thesis in a year and was awarded the degree, though some eyebrows were raised in the psychology department at Manchester because I had conducted no actual experiments. But the head of the department, John Cohen, was my adviser, and he was not of this narrow opinion. (He said to me at the time, "They think if you're not messing about with bits of apparatus then you're not doing psychology!")

Nowadays psychology has pretty much the shape that Chomsky advocated, and it is hard to remember the time when behaviorism was the prevailing orthodoxy. I still think this provides a valuable lesson in questioning orthodoxies that go out of their way to deny obvious facts—as behaviorism in effect denied that we have minds. The sure mark of an ideology, in science and philosophy as in politics, is the denying of obvious facts. A healthy dose of common sense is always a useful antidote to ideological bias.

Before I leave this phase of my life I should mention

three other intellectual events that stand out in my mind from these years. The first is my contact with the Danish philosopher Peter Zinkernagel. John Cohen had met Zinkernagel at a conference and mentioned that he had a student with philosophical interests who might be interested in his work. Zinkernagel was (and is) a neglected philosopher with some striking views, and he was keen to promulgate them. I agreed to read Zinkernagel's *Conditions for Description* and write a commentary on it, which I duly did. Zinkernagel's central contention is that there are rules for using language that cannot be violated without producing a kind of nonsense. Thus there is a rule that prohibits us from using words for experiences independently of the personal pronouns, and there is a rule that prohibits us from using personal pronouns independently of words for physical objects, including human bodies. Accordingly, we cannot make sense of the concept of experiences that no one has, and we cannot make sense of persons that have no bodies. Zinkernagel argues that this refutes idealism and other views that depict the world as a mere collection of ownerless experiences.

These are interesting ideas and I enjoyed thinking about them; and studying Zinkernagel put me in contact with the kind of linguistic philosophy that is associated with Oxford, particularly the work of P. F. Strawson (who would later become another of my advisers). But what I most remember about the Zinkernagel connection was less the substance of it than the contact I had with the man himself, indirect though it was. I sent him my commentary and we began a correspondence that lasted a full year and involved one or two letters a week, sometimes more. No sooner did I send him a letter than he would reply to it, and so on. In my mind's eye I can still see my name and address typed with a worn-out ribbon onto an envelope with a Danish postmark on it. This gave me the experience of protracted philosophical discussion in writing, which was a valuable way of learning how philosophical argument is conducted. It also felt flattering, I admit, to be a mere postgraduate student, not even in philosophy, being taken seriously by an established foreign philosopher. And the internationalism of intellectual life was also brought home to me. I have never met Zinkernagel, nor even seen a photograph of him, but I gather he still lives, still neglected, in Copenhagen, and is now regarded as a bit of an eccentric. When I lectured in Copenhagen many years later he was asked to attend, as I had let it be known that he was a Danish philosopher I had had some contact with; but he never came, and I doubt that I will ever meet him now. The last I heard from him was a few years ago when he wrote to ask me to write him a letter of recommendation for a research grant. I couldn't help reflecting on how things had changed since the days when I used to correspond with him as a callow student from my Manchester bed-sit.

The second episode concerns Wolfe Mays. I had written a thesis, as part of my undergraduate psychology degree, on the history of psychology, focusing on the views of the founders of positivism and phenomenology, namely Ernst Mach and Edmund Husserl. (I also produced an experimental thesis on visual search, as my psychology teachers would never have tolerated a thesis on so "theoretical" a subject as psychological methodology—so I ended up writing two theses instead of the required one.) My theme was that Mach and Husserl held remarkably similar views, emphasizing the primacy of immediate experience, despite the wide divergence in psychological methods to which their philosophies eventually led (behaviorism and phenomenology, respectively). Dr. Mays read this thesis for me, though it was part of my psychology degree and not his proper concern, and he suggested that I convert it into an article that he might be able to publish in a journal he edited, *The Journal for the British Society of Phenomenology*. I did so, and the article was published when I was still only twenty-two. It was my first publication, so I have a certain fondness for it, but it now strikes me as ponderous and overblown. So far as I know it has never been cited by another author. But it did at least give me the idea that I could write and publish philosophical works. There is a special pleasure in seeing one's words in print for the first time; it is as if one has suddenly expanded, doubled. Naturally, I am grateful to Dr. Mays for taking me seriously enough to commission the article. Indeed, looking back on it I find it amazing that he did so. Anyway, thanks.

The third event involved animals. I had met John Harris—fresh out of Oxford himself—who had recently been appointed as a lecturer in the philosophy department at Manchester, because of our mutual interest in Chomsky and innate ideas. As it happens, he was one of three people to edit a groundbreaking book, *Animals, Men, and Morals*, with Stanley and Rosalind Godlovitch. This book lays out the moral case for changing our treatment of animals, and argues specifically for vegetarianism as the only morally defensible position. Peter Singer's celebrated book *Animal Liberation* is a direct descendant of this earlier book, being an expansion of an article he wrote discussing it that was published in the *New York Review of Books*. When the book came out I stood reading it in my local bookshop in Manchester, and by the time I left the shop I was a committed vegetarian. Following Russell, I prided myself on my rationality and susceptibility to a good argument, and I found the case for vegetarianism compelling, so I acted accordingly. The basic argument is simple enough: We should not treat animals in ways that do more harm to them than the good we derive; since killing an animal for food deprives it of far more than we derive from the agreeable taste of its flesh, we should not eat animals for food. Suitably elaborated, this argument persuaded me to change my life in a dramatic and practical way. I even became a vegan for a few years, eating no animal products of any kind (I have had different policies at different times in my life, though I still think the moral case for vegetarianism is conclusive). This was all part of constructing my life according to the dictates of rationality and not blindly following tradition, no matter how deeply entrenched. To me the ultimate sin was refusing to listen to reason.

It had been four short years since leaving Blackpool for Manchester, but the changes in my life had been considerable. I had gone from being an academically under-achieving provincial boy to being close to what I am today—someone whose life revolves around ideas. But there was still a long way to go before I could convert this into a viable way of living. I remember when my parents came to visit me in Manchester soon after I'd graduated and asked me what I planned to do with my life now that I had a university degree. My mother had seen an advertisement in the local paper for a job as a tax inspector, only two A-levels required—would that interest me? I took a deep breath and announced "I want to be a philosopher." Ashen, they asked me to sit down, as they expressed their concerns about whether this might ever lead to paid employment, and indeed wondered just what a philosopher was. I have to admit I felt pretty wobbly myself.

CHAPTER THREE

Logic and Language

OXFORD HAD REJECTED ME IN 1971, THE FIRST TIME I APPLIED, but I hoped to improve my prospects for admission during the year I was writing my MA thesis by working more on philosophy, which I assiduously did. I applied again the following year. But I was rejected again. I had submitted as writing samples my essay on Zinkernagel and my published paper on Mach and Husserl. The first time I had been rejected by the central philosophy panel after being accepted by Balliol College; this time I was also rejected by Balliol. I was nowhere. Evidently, Mach and Husserl, and Peter Zinkernagel, were not revered names in Oxford (this is putting it mildly) or my essays were no good. This was disappointing but not totally surprising, in view of the nature of the Oxford B.Phil as a course for undergraduates who had distinguished themselves in philosophy—and I had only a psychology degree. You might think I should be complaining now about having been done a grave injustice, and so on, but given the realities of admissions procedures that would be an unreasonable complaint. After all, lots of other bright people would have been competing for the limited number of places Oxford could offer.

However, after some negotiations—and protests from my Manchester teachers—I was finally admitted for the B.Litt degree in Oxford, a postgraduate degree (as I later learned) for people who couldn't cut it on the B.Phil, and a distinctly despised two syllables. But I was at least enrolled at Oxford University, where the philosophical action was, and the intellectual glamour. At that time, 1972, Oxford had a large and vibrant faculty; something like seventy philosophy dons were dotted around the various colleges, and it attracted some of the brightest students from around the world, including America and Australia. Oxford had become a—maybe *the*—world center in philosophy in the postwar period, and in 1972 it was still going strong. Moreover, it attracted foreign scholars on sabbatical or giving invited lecture series, so there was a great deal of philosophical activity and excitement about the place. You certainly felt in the thick of things. And, of course, there was an attractive and intoxicating confidence—even arrogance—about the place, born of centuries of distinction and privilege.

I was also entranced by Oxford as a city. After gray and sodden Manchester—with its nineteenth-century industrial atmosphere, its northern functionalism—the ancient colleges of Oxford, with their well-kept gardens and famous "dreaming spires," seemed to supply the perfect environment for serious philosophical study. Walking its streets was an exercise in intellectual osmosis; that an alley should be called "Logic Lane" seemed to me the height of philosophical cool. The college quads seemed bathed in the life of the mind, the very stone heavy with thought. Glimpsed dons were like cerebral gods, remote but real. (Ah, such dreams.) And—no point in denying it—there was the status of the place; being at Oxford meant you had made it, you were somebody, even if you were only enrolled for the lowly B.Litt. I spent the summer vacation of 1972 living in north Oxford, finishing writing my MA thesis, and looking forward to the start of the Oxford term in the autumn. My head buzzed with intellectual anticipation.

I well remember the thrill of seeing the list of classes that was published at the beginning of the autumn term (known in Oxford as Michaelmas term). This was a large broadsheet, like a newspaper, light brown in color, giving details of which classes were to be available, and it brimmed with possibilities. There was just so much to go to, and all of it sounded interesting. Finally, I could immerse myself in philosophy from dawn till dusk. It felt like a liberation after years of having to make philosophy take second place in my studies. Maybe I should have changed subjects sooner, had more confidence in my vocation, but at least I was doing it now. I ticked off about thirty classes I wanted to go to. I felt a slight twinge of regret at saying good-bye to psychology, but philosophy irresistably beckoned.

One of the first classes I attended was given by a young don at University College named Gareth Evans, entitled "Theory of Reference." He would have been about twenty-five at the time, which meant he had stopped being a student only a couple of years earlier. He had longish dark hair, a thin beard, and wore casual clothes (cords, desert boots: This was the early seventies). He was strikingly handsome. He was also, as I quickly discovered, intense and high-powered. To say that he didn't suffer fools gladly would have been an understatement. He spoke with great precision of articulation, in a rather high, penetrating voice, and he breathed philosophical passion. My problem was that I understood hardly anything of what he was saying. He was speaking on a subject of which I had no knowledge, in technical language that was foreign to me. His subject that day was the indeterminacy of translation, as propounded by the Harvard philosopher Willard van Orman Quine, the most distinguished living analytical philosopher of the time. This involved much talk of rabbits and their parts, and the question was how to determine what someone who is speaking an untranslated language is referring to when they say "gavagai" in the presence of rabbits. You might think the obvious interpretation is that they are referring to rabbits when they use the word "gavagai." But, as Quine pointed out, who is to say that they are not eccentrically referring to the parts of rabbits, so that the correct translation of their word "gavagai" is "part of a rabbit"? After all, when rabbits are present so are their parts; so if the foreigners were referring to parts of rabbits, not whole rabbits, then they would still say "gavagai" under the same circumstances. (If you think that part of a rabbit could be present without a whole rabbit, as with a severed rabbit foot, Quine added the proviso that "gavagai" might mean "undetached rabbit part.") Or again, the natives might be referring to a temporal stage of a rabbit, a time slice which lapses while the rabbit persists; or maybe they are referring to those little flies that always hover around rabbits. Nothing in the presented scene will decide which among these competing hypotheses will adhere, and we have no direct insight into what the native speakers might have in their minds with which to settle the question. There is no use in asking them, since we don't yet know how to translate their language. Little did I know it on that day in Gareth Evans's seminar, but this question was one of the big issues being debated in Anglo-American philosophy departments in the early seventies.

What struck me as I listened to Evans was that this seemingly trivial question had major ramifications: If we could not resolve the indeterminacy of reference of the term "gavagai," then maybe the whole idea that we refer to discrete things in the external world was a kind of illusion. Indeed, Quine argued just that: There is, he insisted, just no "fact of the matter" about what people refer to with their words; you can assign alternative schemes of reference to a language, even your own, and not be accused of error. Reference is a kind of myth, a prescientific holdover, according to Quine. And yet doesn't it seem clear that I at least *think* I am referring to rabbits when I say "rabbit"? But this objection would not impress Quine, who would dismiss such a response as unverifiable "mentalistic semantics" ungrounded in observable facts of behavior. Quine wanted a semantics that depended only upon objective, publicly observable facts of linguistic use, not ethereal "ideas" hidden away in people's supposed minds. In Evans's seminar I saw rigor and clarity, and a high level of technical sophistication, brought to a difficult and profound question. It was my first taste of contemporary analytic philosophy, and my appetite was piqued. If only I knew what all those impressive logical-sounding words meant!

I also got a first taste of the cut and thrust of philosophical debate. Evans was a fierce debater, impatient and uncompromising; as I remarked, he skewered fools gladly (perhaps too gladly). The atmosphere in his class was intimidating and thrilling at the same time. As I was to learn later, this is fairly characteristic of philosophical debate. It is not the sonorous recitation of vague profundities, but a clashing of analytically honed intellects, with pulsing egos attached to them. In fact, truth to tell, philosophy and ego are never very far apart. Philosophical discussion can be a kind of intellectual blood sport, in which egos get bruised and buckled, even impaled. I have seen people white and dry-mouthed before giving a talk to a tough-minded audience, and visibly shaken afterward. No one likes to be publicly refuted, and in philosophy it happens all the time. In Evans I saw someone with considerable debating skills, and I was no doubt attracted to the kind of power and respect that goes with that. Plain showing off is also a feature of philosophical life.

It quickly became apparent to me that I was way behind in the kind of expertise taken for granted in the world of Oxford postgraduate philosophy. I hadn't even heard of many of the figures whose works were well known to my fellow students: Donald Davidson, Saul Kripke, David Lewis, Michael Dummett, Hilary Putnam, and many others (I shall say more about these philosophers as I proceed). I also didn't understand much of the terminology used to discuss these figures: "modal operator," "counterfactual," "predicate extension," "second-level function," " existential quantifier," "scope distinction," "*de re* necessity," and so on and on. I had a lot of catching up to do, clearly.

My first adviser at Oxford was Michael Ayers at Wadham College. I had read an article of his that sounded a bit like Zinkernagel, so I asked Dr. Ayers if he would take me on. He was a tall, thin man in his thirties, with heavily creased pouches under his eyes, and was obviously a product of Cambridge (serious but with a hearty laugh, and no airs—despite his phonetically identical surname). We met once a week in his small room in Wadham to discuss whatever paper I had written for him that week. Zinkernagel soon went by the board and I began to work on the topics that dominated debate in Oxford, mainly logic and language. Under Dr. Ayers's friendly and probing supervision I made good progress, and at the end of the first term he recommended to the philosophy panel that I be permitted to enter for the B.Phil. degree. I therefore met with Professor R. M. Hare, White's Professor of Moral Philosophy, who had shunted me onto the B.Litt earlier, and he seemed gratified that I was now working on topics like proper names and counterfactuals—not phenomenology and obscure Danes. He agreed to transfer me to the B.Phil. This made me happy, but I also knew that it wasn't going to be easy making the grade. I was surrounded by some very bright and well-educated philosophy students, much further along in their studies than I. I had already seen some incoming B.Phil students wilt from the intensity of the intellectual competition—now suddenly thrashing around in a bigger pond than they were used to—but at least I didn't expect too much of myself, given my educational background. I was justifiably modest about my philosophical standing.

Among the most noted students was a young man named Christopher Peacocke. He came from an academic background and had been educated at Oxford as an undergraduate. He had won all the prizes for being the best student in his year as an undergraduate, and he certainly knew his stuff. He spoke the logical jargon as if it were his mother tongue, and operated on a highly abstract level. His mouth hardly moved as he constructed sentences like "The theory of the actual language relation needs to supplement a Tarskian truth definition with an account of S-meaning that respects the constraints from the recursive structure of language." He wore thick glasses and dressed in a jacket and tie, even while sitting in his rooms alone. He brimmed with philosophical enthusiasm. We used to meet for tea and discuss what we were reading; as he covered napkins with logical formulae, I tried to get things clear. These conversations helped me develop my grasp on the material I was studying and hearing about, and I enjoyed discussing philosophy with someone as keen and sophisticated as Chris. Sadly, we were to fall out in later years, when I regretfully came to the view that professional rivalry was more important to him than friendship, but these early conversations played their part in pushing me in the direction I needed to go in. There is a period in one's twenties when philosophical discussion with a friend can be the most delightful of experiences, and an ideal way to make progress in one's thinking. And it helps relieve the solitariness that is so much a part of the contemplative life.

During my time at Oxford two famous American philosophers came visiting: Saul Kripke and Donald Davidson. Their work had been causing a great stir in Oxford and elsewhere, and it was essential to master their ideas if one was to be taken seriously at all. Let me then spend some time describing the outlines of their views. Both were working mainly in the philosophy of language, and their work was demanding and dense, especially Davidson's. Kripke had been a child prodigy, publishing important work in modal logic (the logic of necessity and possibility) in the *Journal of Symbolic Logic* in his teens, and he was still only in his early thirties when he came to Oxford to deliver the prestigious John Locke lectures. He had a tendency to rock to and fro all the time and was a notoriously messy eater, a dispenser of crumbs to his neighbors at the

table. There are many anecdotes about Kripke's eccen-
tricities, but the one I like the best goes as follows. At a
conference a group of philosophers were playing guitars
and singing folk songs after the formal sessions were over.
They asked Kripke to join in and he replied, "If anyone
else did, that would be the end of it, but if I do, it will be
just another Kripke anecdote." (This is what we philosophers call technically a "meta anecdote.") Kripke was a
spellbinding lecturer, with a distinctive nasal voice that
lurched from low to high in unpredictable bursts, and he
seldom stooped to using notes; his most influential work,
Naming and Necessity, is in fact a transcript of lectures
he gave without notes to an audience at Princeton in
1971. Surprisingly, what really got people's attention was
Kripke's theory of proper names such as "Plato" and
"Richard Nixon," a seemingly trivial subject. It almost
sounds like a parody of analytic philosophy to say that
one of the most influential thinkers in the analytical tradition made his name with a theory about proper names.
But once you understand the significance of the issue it is
really not surprising at all.

When you use a name like "Plato" you refer to a certain long-dead historical individual; you talk *about* that
ancient Greek philosopher and make remarks that are
true or false of him. A relation is set up between your
words and that person from the past. But how do you do

that? What makes "Plato," as you use it, the name of that particular individual? It's not as though the name "Plato" somehow resembles the man Plato. You make a sound by uttering the name, and a relation of denotation is set up between you and the ancient philosopher. But what is it that sets up that relation? Why are you referring to one person and not another when you make that sound? To this question there is a natural answer, favored by Gottlob Frege and Bertrand Russell, which had become orthodox until Kripke came along: When I use "Plato" to refer to Plato I have in my mind an idea of Plato, and this idea consists of the various descriptions I would apply to him, such as "the author of *The Republic*" or "the best-known pupil of Socrates." If we go on to ask how "the author of *The Republic*" refers to Plato the answer is straightforward: Plato is the man who *fits* this description, of whom it is true; the description refers "descriptively," by specifying a property that Plato uniquely satisfies. So, if the name is simply short for the description, then the name refers descriptively too: "Plato" just *means* "the author of *The Republic*," and we know quite well how the latter term refers. True, the name doesn't *look* like a description; but it is, the orthodoxy maintained—only a disguised description. We use names as short versions of descriptions, thereby saving our breath. And this is the key to reference in general: We refer to things in the world by describing them in certain ways, implicitly or explicitly. This theory is called the "description" theory of reference. To many it seemed self-evidently true, the only way reference *could* work. How do we refer to things? By knowing what properties they have and identifying them via those properties. Simple.

But Kripke saw problems with the description theory (as did some other philosophers at around the same time). Can't we refer to Plato with the name "Plato" and not know a true description of him that distinguishes him from every other person? After all, many people have used the name "Plato" and not known that he authored *The Republic*, or any other fact about him that singles him out; a user of the name may have just thought, "Yes, Plato, heard of him, some ancient Greek who wrote a lot of books." If that's all you know about Plato it wouldn't stop your referring to him by name, would it? Moreover, we sometimes make mistakes about the people we name. If I think of Jerry Lee Lewis as "the guy who invented rock 'n' roll" I can still refer to him, even though it was in fact Little Richard who invented rock 'n' roll. It's not that when I utter the name "Jerry Lee Lewis" I thereby refer to Little Richard, just because I mistakenly believe that Jerry Lee invented rock 'n' roll, while in fact it was Little Richard who has this honor. So what ties the name to the person named cannot be the descriptive information I associate with the name; there must be something else at work. The description theory forgets that we sometimes have false or sketchy beliefs about the people we name, so that our beliefs cannot be relied on to pick our reference out uniquely.

What might be put in the place of the description theory? Here Kripke suggests that what forges the link is a line of historical and causal connections between uses of the name and the person named. A baby is born and baptized; the name goes into family use; it spreads out to others who are connected to these initial users of the name; the named person eventually dies while the name lives on, as each speaker in the chain picks up the name from someone he or she heard it from. This is how we refer to Plato with "Plato": There is a chain of uses of the name stretching back to the speakers of Greek who first referred to the baby Plato. So long as you are a link in this chain you can refer to Plato, even if you know little about him and may even have false beliefs about him. The mechanism or process that connects the name to the person consists of a vast range of social, historical, and causal relationships, which occur outside the mind of the speaker of the name. It is not that I have to *know* all about the vast chain that links me linguistically to Plato; it is enough that the chain exists and that I am a link in it. All I need to do is intend to use the name in the same way as the person I heard it from; the rest is accomplished by the existence of the chain itself. To change the metaphor: Each speaker is a node in a network, and reference works because the network is causally linked to the objects in the world that we speak about.

This picture of naming contains three important elements that changed the way philosophers of language thought about reference and led to what is now called The New Theory of Reference. First, it introduced a social element into the functioning of language: It is not that each speaker is linguistically on her own, with only her own mental resources to enable reference to occur; rather, she is socially connected to others who may know far more than she does about the reference of the terms she uses (this came to be called "the division of linguistic labor"). Second, and relatedly, the mechanism of reference is now not located inside the mind of the individual speaker, as it was with the description theory, but consists of the way the speaker is embedded in the wider world—what she is externally connected to. Third, there is a causal element in reference: There is a causal chain leading from the object to a use of the name for that object, and this causal chain is what reference ultimately depends on. This element led to what later came to be called the Causal Theory of Reference—the idea that reference could be analyzed as a special kind of causal relationship between words and the things in the world that words refer to. And this in turn led to the idea that it might be possible to "naturalize" reference—to show that it could be slotted into a scientific account of human linguistic activity. If reference is just a fancy kind of causal relation, then it is no more mysterious in principle than any other kind of causal relation—for example, the collision of billiard balls. So what looked at first like a trivial issue about proper names turns out to involve big questions about how language hooks up to the world and ultimately how thought manages to be about things (more on this later).

The writings of Donald Davidson were also attracting something of a cult following in Oxford in the early seventies (people spoke of a "Davidsonic boom"). At this time Davidson had published only a handful of short, densely argued articles, mainly in obscure places; terse, technical, and forbidding, these articles were a kind of IQ test for aspiring philosophy graduate students. Davidson himself, with his piercing blue eyes, clipped speech, and confident manner, was also visiting Oxford and shaking things up with his new theory of meaning. Davidson always managed to give the impression that he had it all figured out, if only you would pay close enough attention. It was all just a matter of shedding the right misconceptions and prejudices; then everything would be crystal-clear, pebble-smooth. Philosophers have long asked "What is meaning?"or "What do we know when we know the meaning of a sentence?" Meaning seems elusive, ethereal: Where have you ever seen a meaning, or stepped on one? Some, like Quine, were openly hostile to it, wishing it would go away, calling it nasty names ("museum myth," "creature of darkness"). Davidson proposed that we ask what a *theory* of meaning ought to be like: What should such a theory do and how should it be constructed? Well, it should specify the meaning of every sentence in the language that has a meaning. And it should do this by meeting two "constraints" (how well I remember the way Oxford philosophers loved to use the word "constraint"): First, the theory should specify the meaning of every sentence by reference to the meaning of the words that make up the sentence; second, the theory should do this without already assuming the notion of meaning, or else it would be unexplanatory and circular. The question then became how to specify the meaning of sentences in a noncircular way, and how to derive this specification from the words that make up the sentence. How, to choose the favorite example, do we say what "snow is white" means, in such a way that we exhibit this meaning as resulting from the meaning of the words in the sentence? And if we can answer this question, won't we have solved the age-old philosophical problem of meaning? Isn't meaning whatever a good theory of meaning is a theory of, so that the theory ought to contain an account of the very nature of meaning?

Davidson's "bold hypothesis" consisted of two parts: The meaning of a sentence is its truth condition, and truth conditions could be specified by using the logical machinery supplied by Tarski's theory of truth (Tarski was a Polish logician who established many important results in mathematical logic). The sentence "snow is white" is true if and only if snow is white, and the same holds for the sentences of other languages that say the same thing. When we understand a sentence such as this, what we know is its truth condition, what would make it true—so meaning and truth conditions seem intimately connected. If I tell you that the Italian sentence *"Il neve é bianco"* is true on condition that snow is white, you thereby learn the meaning of this sentence. So meaning is contained in truth conditions; moreover, the concept of truth does not presuppose the concept of meaning, so we seem to be meeting our noncircularity "constraint." Now comes the hard part: Tarski's theory of truth. This is actually a very technical theory in its details, but the central idea is quite simple. By virtue of what does the sentence "snow is white" have the truth condition it has? Clearly, the words "snow" and "white" have everything to do with this, but what is it about these words that determines how they contribute to truth conditions? Well, "snow" refers to snow and "is white" is true of something if and only if that thing is white; if we put together these two pieces of information we can derive the result that "snow is white" is true if and only if snow is white. We analyze the sentence into two parts, subject and predicate, corresponding here to "snow" and "white," and then we specify a reference for the subject term and a condition of satisfaction for the predicate term; the sentence is then true if and only if the reference of the subject term satisfies the condition of satisfaction associated with the predicate term. The difficult technical part of Tarski's theory of truth lies in the details of how to do this for all the types of sentences there are, not just simple subject-predicate sentences; and this part I will leave for my reader's homework (to be completed by the beginning of 2010, please). The key idea of Davidson's theory, then, is to specify truth conditions on the basis of the parts of sentences, by saying something relevant about the parts. That will be our theory of meaning.

The reason this seemed so attractive was that it promised to answer the old, obscure question "What is meaning?" with a nice new technical theory. Now we could talk about axioms and theorems, recursive clauses, Convention T: We had made mathematics out of philosophy. The philosophy of language could now be put on the road to scientific respectability. Philosophers like

Wittgenstein (about whom more later) had ruminated about meaning, speaking of linguistic use as the key to meaning, describing different language games, finding deep puzzles in the nature of meaning, but all this could be swept away in the bright light of logic and mathematics. The issues became purely technical, a mere matter of writing your axioms the right way to get out the theorems you were looking for. It was the ever-tempting hope of turning philosophy into science—misguided, perhaps, but undeniably appealing ("sexy," as some philosophers like to say).

I wrote my B.Phil thesis on this kind of "Davidsonian semantics," focusing on a troublesome question for the approach: how to handle sentences that have no truth conditions. It can hardly be that when we understand "Shut the door!" we know its truth condition, because this sentence cannot be said to be true or false. Rather, it can be said to have been complied with or not complied with. Similarly for "Is the door shut?" and "Would that the door were shut!" My suggestion was to construct theories with the analogue of truth conditions, such as obedience conditions. The sentence "Shut the door!" is obeyed if and only if the person addressed shuts the door. There were some niggling problems with these so-called non-indicative sentences, but something along the lines I suggested seemed the right way to go. I wrote out my axioms, specified my logical forms. It was a far cry from Sartre and the ontological argument and free will. Philosophy for me had morphed into logic and language, under the encouragement of Oxford. I still think that any professional philosopher should have a good mastery of these logical and linguistic ideas, but I no longer believe that these ideas alone will lead to the resolution of serious philosophical problems. To that extent, then, I do not believe that philosophy can be a science. But I do understand why my twenty-three-year-old self had a hankering for this to be possible: Reducing the mystery of the world by scientific methods is certainly an attractive prospect, the main intellectual force that has shaped the modern world. Maybe, too, doing philosophy in this technical style is a way to increase one's intellectual prestige in a world that values scientific knowledge above all else—not to mention the just plain showing off that technical expertise permits. In any case, it was fun while it lasted.

During my second year on the B.Phil I had the benefit of Professor P. F. Strawson as my adviser, now elevated to Sir Peter Frederick Strawson. He was quietly skeptical of the scientism that we graduate students were imbibing, encouraging me not to lose sight of the traditional philosophical issues; I now see what good advice this was. I would meet "P.F.S." once a week in his rooms at Magdalene College, where we would talk about whatever philosophical issues were on my mind. His gentle probing, clarity of thought, and powerful mind were an important stimulus to me. He was (and is) a man who is austerely elegant in all things: in his writings, certainly, but also in his dress, the way he entwined one leg around the other while discussing philosophy, and the Wildean manner in which he smoked a cigarette. I remember when I first went to see him and he asked how my formal logic was; apologetically, I said it wasn't all that advanced; he smiled and replied "Good!" fearing I would be one of those tiresome inelegant technocrats trying to suck the spirit out of traditional philosophy.

Another important influence was A. J. Ayer, aka Professor Sir Alfred Ayer, aka Freddie Ayer to his friends. Professor Ayer was the most publicly famous philosopher in Oxford, a regular TV "personality," and a contributor to the Sunday papers, among much else. He lived in London and lent an air of worldly metropolitan sophistication to the somewhat precious hothouse atmosphere of Oxford philosophy. I began attending his "Informal Instruction" classes, in which a book would be chosen for discussion for the term, to be introduced by a student and thereafter conducted by Ayer in his pacing rapid-fire style, a cigarette permanently afire. At the first session Ayer asked who had read the book to be discussed—*The Nature of Things* by Anthony Quinton. I happened to have just finished reading it, so I raised my hand; to my surprise no other hand went up, and a cold shiver went through me as Ayer fixed me with a beady eye. This was an intimidating gathering for the lad from Manchester, and I was afraid he would ask me a question about what I had read. I had always been nervous in contexts of public discussion and feared that the words would literally not come out of my mouth if I were put on the spot. Little did I know what lay in store for me. The session continued, with Ayer giving the first presentation of the term, followed by what seemed to me like a very high-powered discussion, to which I did not even think of making a contribution. At the end of the class, however, Professor Ayer suddenly fixed gaze his on me, hunched at the back of the room, and announced, "The man at the back can pay for his virtue and give the presentation on chapter two for next week." He didn't even wait to see if I was agreeable to this brilliant suggestion. That was it: me, next week. The subject was identity through time, which was quite new to me, and I had a week to get something together that would not be a complete disgrace. I felt the blood drain abruptly from my face.

That week I worked with a nervous agitation I had never experienced before. I consulted my then adviser, Michael Ayers, on what to read, and we had a discussion about the topic. I wrote about ten pages, hoping at least to show that I was not a complete dunce. I dreaded the arrival of Wednesday, when I would have to go before the firing squad, counting the minutes up to 5 P.M. when the class started. I duly found myself in front of about forty clever people, ready to find fault with whatever I had to say. I read my essay aloud, staring self-protectively down at the page. When I had finished, I looked up, as red as a beetroot, with very clammy palms (which I always get when I am nervous), and Professor Ayer said, matter-of-factly, "Very good," and launched into his own view of the subject. I think he was happy that I had mentioned Hume, his favorite philosopher, and the simple-mindedness of my presentation probably impressed him as confident clarity. I sat in a bit of a daze for the rest of the session, trying to avoid any further engagement, but I did get drawn in a few times and found that my vocal organs could function after all. At the end I was dizzy with relief, and a few of the other students came up to continue the discussion, evidently not thinking me a hopeless case. It was the first time I had presented a philosophy paper, a rite of passage for anyone hoping to become a professional philosopher—and I had survived the ordeal.

However, that hope was still very far from being fulfilled. Academic jobs were exceedingly scarce in England in the early seventies, after the boom years of the sixties, and there were lots of very able people seeking them. All I had was a degree in psychology from Manchester, not a philosophy degree from a prestigious university. I would need to obtain a job offer before being awarded my B.Phil, assuming I could pass, in order to be able to start teaching in the following autumn, when my grant would run out. It looked pretty hopeless, and I had no clear plan. In order to improve my chances on the B.Phil I decided to enter for a voluntary examination called the John Locke Prize. This examination is for people aiming to win the prize of one hundred and fifty pounds, along with the prestige that goes with it. Traditionally, the brightest philosophy post-graduate at Oxford wins the prize, though there are famous cases of brilliant people not being deemed of sufficiently high standard by their examiners to be worthy of the prize. Christopher Peacocke had won it the year before, not to anyone's surprise. The John Locke Prize is not always awarded in any given year; it does not automatically go to the best person entering in that year. It goes only to someone deemed to be of a sufficiently high standard, and there can be years in which the examiners do not think that anyone has met their very demanding requirements.

My purpose in entering for this examination was simply to get some practice taking philosophy examinations, since I had not done any before (not counting Sartre). The examination is conducted over two days, with four papers, each three hours long, with three questions to be answered per paper. It thus resembles the B.Phil examination, affording me a good dry run for the B.Phil itself (which I knew to be regularly failed by a few people each year). The examiners that year, 1972, for the John Locke Prize were Professor Ayer, Professor Hare (who had let me onto the B.Phil), and Brian Farrell, the Wilde Reader in Mental Philosophy—a fairly formidable crew.

I turned up in "subfusc" for the first examination: white bowtie and shirt, dark suit, black shoes, cap and gown (this was compulsory: anything missing and you would be denied entry to the examination hall). I buckled down to the questions, writing about logical form, the coherence theory of truth, the nature of necessity, personal identity, the analysis of knowledge, and so on. About halfway through I very nearly walked out: I was struggling with some of the questions, and why put myself through this agony? But I persisted, telling myself I needed the experience. I finished all four papers and left feeling exhausted and severely tested. There were about ten candidates, all voluntarily submitting themselves to this labor in the hope of eventual glory. It can be grueling stuff, this philosophy.

About a week later Professor Ayer informed me that my handwriting was so bad that I would need to have my papers typed by a professional typist in the presence of an invigilator to make sure I hadn't cheated. Moreover, I would personally have to pay for this to be done. I expressed my misgivings, saying I had not acquitted myself at all well, and worried about the enormous expense of about fifty pounds that this was going to cost me. Ayer replied that I, or it, was "worth it," so I reluctantly agreed—and anyway, you didn't not do as Sir Alfred told you to do. I accordingly read my atrociously written papers aloud to a bored typist in the presence of an equally bored invigilator, who awoke to take exception to my inelegant use of the phrase "chunk of reality," wincing all the way. I really must improve my handwriting, I thought. (Even today my writing is a miracle of illegibility.)

Then a week or so later, as I was sitting down for one of Kripke's John Locke lectures, Professor Ayer conspicuously approached me in front of about five hundred people, clapped me on the back, and told me I had won the John Locke Prize—and by a wide margin. It must have been obvious to many of the people there what Ayer was communicating to me. He then congratulated himself, characteristically, on his good judgment in having ordered my papers typed and strode off to regain his seat. Kripke's lecture started and I sat down in a daze of amazement and elation; I don't think I heard a single word that Kripke said for the next hour. In that brief moment my entire life had turned around, because now I was in a favorable position to obtain an academic appointment in philosophy; such was the prestige of the John Locke Prize. I was within reach of making my dream of becoming a professional philosopher a reality. Professionally, it was the most momentous day of my life, and well illustrates how it is possible to beat the odds and become what you yearn to be. As it turned out, two other candidates were also deemed of sufficient standard to win the prize that year, one of whom had already won the prize for the best undergraduate papers in Oxford finals. I believe this was the only time that three candidates were posted as worthy to win, with me at the top.

Of course, my reputation shot up dramatically after that and I became a minicelebrity in the world of postgraduate philosophy at Oxford. Looking back, I think what had happened was that my relative inexperience worked in my favor, because I had to struggle to create my own answers to the questions, for lack of knowledge of what others had said. Instead of jamming my answers with too much secondhand knowledge of the philosophical literature, I was forced to work through a line of thought of my own, thus demonstrating an ability to engage in original philosophical reasoning. I confess that at the time part of me thought it was a complete fluke and that the B.Phil examinations would reveal this. But I obtained a distinction in these too, when the time came, so I conceded to that doubting part of myself that I must have something going for me. The other half of me—the half that let winning the prize go (oh yes!) to my head—thrilled to the thought that I must be a hell of a good philosopher to do so well after so little time. Whatever may be the case, it was certainly just what I needed. I wonder now what would have happened to me if Ayer had never asked me to have my papers typed (a highly unusual step, in fact), or if I had walked out when I felt like it or if I had just not sat for the John Locke Prize at all. Things would undoubtedly have been very different, and even now I feel a cold sweat at the alternative possibilities. Life and chance, chance and life.

Both my advisers, Michael Ayers and Peter Strawson, were fulsome in their congratulations, but Professor Hare, White's Professor of Moral Philosophy, had an odd reaction. You will recall that it was he who initially declined to allow me in on the B.Phil and who later agreed that I be upgraded. He never congratulated me on the prize, and indeed seemed to avoid me when our paths crossed. I assume he felt embarrassed at his lack of insight in not spotting me for a decent philosophy student a year earlier—though I didn't myself think there was any reason for embarrassment on that score. He had made a perfectly rational decision earlier; it just transpired that I did better than could have been expected. The thing to do was to say to me something like: "Well, that's quite a turn up for the books!" As it was I found his embarrassment and inability to acknowledge what had happened rather weak—and not what I would have expected from someone in his position as a distinguished moral philosopher. Perhaps it was just a matter of ineptitude in personal relations, but I felt it to be somewhat of a slight, caused by an inability to admit when you have been wrong (even though his earlier opinion was quite justified at the time). Other than that winning the prize was unalloyed joy.

I left Oxford in 1974, after being there for just two years. In that time I went from the periphery to the mainstream of philosophy, making the transition to a world that is still pretty much the one I occupy today (give or take a continent). It was a formative and decisive time for me. I would have liked to stay in Oxford for longer, maybe as a junior research fellow for a couple of years; I had no illusions that I might be able to get a real teaching job there. In the end I applied for a job at University College London (UCL) and was interviewed there. I didn't get the job I applied for— that went to Arnold Zuboff from Princeton—but another job became available during the interviews, owing to the early retirement of one of the lecturers there, and I did get that job, no doubt largely because of the John Locke Prize. I was scheduled to begin teaching in the autumn of 1974, which meant that I would suffer no serious financial lacuna. I felt very lucky to obtain that job considering the shortage of positions for philosophers, but I was sorry to leave Oxford, which I felt to be the center of philosophy in England, if not the world. Still, London and Oxford are only an hour's train ride apart, and I knew I'd be back. I was only twenty-four (and looked even younger), and still quite undeveloped philosophically (and otherwise). At the time of being appointed at UCL I had no formal qualification in philosophy, since I had not yet taken my B.Phil examinations and had no undergraduate philosophy degree. I had gone from scraping into Oxford from a background in psychology to obtaining a job at one of the best philosophy departments in the country. When I look back on this it strikes me as a kind of miracle, a series of immensely lucky breaks—though of course I had worked hard during my time at Oxford, day and night. My interests had become technical, focusing on language and logic, and I had lost all interest in the mind—possibly as a reaction against psychology. I was perhaps a little priggish, certainly serious about my work, not very worldly. My life was thoroughly consumed by philosophy, which I used to continue reading even in bed at night. I liked nothing better than a good long philosophical argument. This intellectual urge was strong and deep in me, almost exclusively so. I couldn't quite believe where I had come, but I was determined to make the most of it.

CHAPTER FOUR

Mind
and Reality

AS A POSTGRADUATE I LIVED IN A NICE COLLEGE FLAT IN POSH
north Oxford, which I rented from Jesus College for the
princely sum of eight pounds a week; I even had a study
looking out over a garden. Lady Somebody lived below me
on the ground floor (we had some chatty teas together,
she and I). I could easily cycle from home into the center
of town. I was very short of money, but I needed little in
order to live a fairly agreeable life. In London I went back
to living with a landlady, a Mrs. Collins, a round and red-
faced woman who had fallen on impecunious times but
was cheerily accommodating, renting me a small room in
a large flat near the Albert Hall in Kensington. I had no
phone and, worse, no desk. It was a forty-minute trip on
the underground to Bloomsbury, where the UCL philosophy department was located, with a change of trains on

the way. My starting salary was derisory even by academic standards. My first job had made me worse off.

The main part of my job was teaching tutorials—tutorials, tutorials, and more tutorials. A group of students—in threes, twos, or ones—would come to my office once a week and talk about an assigned philosophical subject. I usually had about eight groups a week. Then there were lectures and seminars to give, as well as graduate supervision, and department meetings. My initial enthusiasm for teaching dampened after about the hundredth tutorial, a little over a term's worth. It is hard to convey the particular fatigue that sets in during the third tutorial of the afternoon, especially after you have given a lecture in the morning and worked on your own stuff for a couple of hours. You have to pay careful attention to the half-formed thoughts of the students, who often endeavor mightily to sit in abject silence for the hour, and to respond as helpfully as possible to their problems. Instead of giving a lecture, in which you set the pace and tone, you have to go at the student's speed and address his or her particular confusions and difficulties. By the end of the day your head feels as if it's about to implode, a deflated soccer ball of botched kicks. And yet there is that obligation to do the best for your struggling students, no matter how tedious it may be. Before long I was gritting my teeth as I climbed the stairs to my office, lingering

wistfully before the last flight, knowing that I had an especially mute and tin-eared group to contend with.

My office itself was a cramped space on the fifth floor
that led to the fire escape on the roof, with no lock on the
door, and very cold in winter (there was a tiny gas fire in
the corner). The absence of a lock proved a problem when
one of my female students stole several of my books,
including an embossed one that Jesus College had given
to me in honor of winning the John Locke Prize (I never
got it back). There I would sit, hour after hour, squeezed
close to the tutorial students of the moment, trying to
keep up my concentration. Oh, Socrates, was it thus for
you? Not that tutorials were all fatigue and frustration: I
did have a number of very bright and enthusiastic students who would do the reading and come along with
something intelligent to say. But far too many of them
came ill-prepared and indolent, expecting me to do all the
work. And when you are explaining Russell's Theory of
Descriptions for the thousandth time—for the third time
that very day—encountering the same misunderstand-
ings over and over again, it is hard to keep up much interest, no matter how able the student. No doubt tutorials
are good for the student—though their virtues are often
exaggerated—but they can be one of the outer circles of
hell for the teacher. Imagine being forced to teach the
same tutorial over and over again for all eternity! (You

may think I'm exaggerating; I am, slightly, but I want to get the point across as strongly as possible—and I have no doubt that the philosophy tutors among my readers will be dolefully nodding their aching heads at this moment.)

My first lecture course was on the topic of truth. The question was: What is the meaning of the word "true"? Or: What is it for a statement or belief to be true? Some philosophers have held that a given belief is true when it coheres with the other beliefs that you hold—when it is part of a consistent web of belief. The trouble with this is that coherence or consistency cannot be sufficient for truth, since these relationships among beliefs are compatible with their falsehood: The belief that grass is blue and the belief that snow is purple are perfectly consistent with each other, but neither of these beliefs is true. Beliefs have to fit the world to be true, not merely fit each other. This leads to the so-called Correspondence Theory: A belief is true when it corresponds to the facts. That sounds like it's on the right lines, but it is rather obscure what this relation of correspondence is supposed to be, and the mention of facts raises questions about what sorts of entities facts are. For these reasons, among others, many twentieth-century philosophers have favored the Redundancy Theory of truth—and this was the view I defended in my lectures. To say that a belief is true is simply to say that things are as they are believed to be. To say that the belief that snow is white is true would simply be to say that snow is white; the word "true" exists in our language merely to avoid repetition—strictly speaking, it is redundant. To explain the concept of truth we therefore do not need to appeal to obscure relations of correspondence and dubious entities called facts; we can capture the way truth depends on fitting the world simply by noting that to call a belief true is nothing more than an indirect way of making a statement about the world. If I say that your belief that smoking causes cancer is true I am simply agreeing with you that smoking causes cancer. The apparently deep philosophical problem of truth vanishes, once we notice the actual function of the word in our language: "True" is a word we use to talk indirectly about the world, when we find it inconvenient or infelicitous to speak directly about it. It is not some mysterious, inscrutable relation that beliefs bear to facts, whatever they may be. For a statement to be true is just for the world to be as it is stated to be—period, argument over.

I still think this theory of truth is basically correct, though I would need to qualify it a bit now. It certainly appealed to the no-nonsense cast of my mind at this time. We had Tarski's formal theory and the redundancy theory, and together they told the whole truth about truth. That was my message to the students: Truth has been tamed, deflated, demystified. During the course the head of the philosophy department, Richard Wollheim, whose main interest was psychoanalysis, attended one of my lectures, in order to monitor my lecturing skills. "Very good, Colin," he said, in his mellifluous voice, "but you follow the method school of lecturing—you mumble your lines and keep your back to the audience at all times." The fact is that I was so taken up with articulating the ideas that I gave no real thought to my manner of delivery. I resolved to work harder on my lecturing style—and you will be glad to hear that within a matter of only a few short years I was managing to face the students for most of the lecture period. I have never been much of a performer, and I try to let my clarity compensate for my lack of showmanship.

I was very interested at this time in the concept of necessity, stimulated partly by Kripke's groundbreaking work. Among the things that are true, some of them are true in an especially strong way—they are necessarily true. These truths differ from the truths that could have been otherwise—those that are only contingently or accidentally true. For example, it is true that I am a philosopher, but that is not a necessary truth: It is perfectly conceivable that I could have gone into some other field—psychologist, drummer, plumber. Most truths are like that: They just happen to be true, but they are not true "in all possible worlds." I am drinking decaffeinated coffee now, but that is hardly something that could not conceivably have been otherwise. Contrast this kind of case with something like the number two being an even number or bachelors being unmarried or everything being identical to itself. When you think about it you see that these facts could not have been otherwise: The number two could not have been odd and still be the number two; bachelors are by definition unmarried, so there is no sense in the concept of a bachelor who is wed; an object could not fail to be identical with itself, no matter how peculiar it might otherwise be. These, then, are necessary truths—they hold true "in all possible worlds." There is a steadfastness to their truth, an inability to be anything other than true. Even God could not make them false.

Now, when we ask what the necessity of a necessary truth depends on it is natural to give the following answer: It results from our concepts, from our language, from our ways of describing things. Thus it is part of the *concept* of the number two that it be even; it is written into the *meaning* of the word "bachelor" that bachelors are not married; it is simply part of the very *idea* of an object that an object is self-identical. In a sense these seem like trivial tautologies, just like saying that a triangle has three angles. We know these necessary truths just by knowing our language; it is not that they reflect some fact about reality that exists independently of language. The technical word for this characteristic is "analytic": These necessary truths appear to be analytic in the sense that they depend purely upon the analysis of the meanings of the terms involved, not upon anything that lies outside of language. The contingent truths, by contrast, do depend upon facts outside of language: What makes it true that I am drinking decaffeinated coffee is not simply the meanings of the terms "coffee," "drink," and "I," but the fact that the world happens to be a certain way. You cannot, then, ever know that a contingent truth is true just by analyzing the meanings of the words that make it up; you have to have a look at how the world is actually constituted out there. So it is very natural to believe—and so says a dominant philosophical tradition, the tradition called *empiricism*—that the necessary truths are all analytic, matters of linguistic definition, while the contingent truths are all "synthetic," meaning that they are not a matter of definition but depend upon facts that go beyond language.

However, Kripke questioned this neat picture, arguing that some necessary truths hold independently of linguistic meanings. His most controversial example involves one's family ancestors. Each of us was born to a particular pair of biological parents—in my case Joe and June. Clearly this is not merely a matter of the meaning of the name "Colin McGinn": Lots of people who know my name do not know who my parents are, so it is not possible to analyze my name and figure out my ancestry. But when you reflect on it we seem to be dealing with a necessary truth here: I *could* not have been born to anyone except Joe and June. We have to be careful in interpreting this claim correctly: Of course it is possible that there has been some mistake, that I was really born to Jack and Ethel, and that Joe and June snatched me from the hospital claiming me as their own—stranger things have happened. But *given* that it is really true that Joe and June are my parents, that does not seem like the kind of fact that could have been otherwise. I could have been born in a different hospital or at home, without this compromising my identity, but there is no real possibility of my being born to different parents and still being me. I am linked to my parents by the bonds of necessity, and this necessity does not arise from language. It is a necessity written into the nature of things—a "*de re* necessity," as the jargon has it. It is just a fact about me that I could not have had different parents, and this has nothing to do with the meaning of my name. Thus reality itself can have so-called essential properties. Objects give rise to necessities independently of language. Even if there were no language it would still be a necessary truth that I have my origin in Joe and June. Not all necessity is analytic. This demolished the empiricist tradition that located all necessity in language.

My first two published papers, after coming to UCL, were on the topic of essential properties, written during my first term there. One of them was about such properties of gold and other "natural kinds": here I argued that not only is the molecular structure of gold an essential property of it but so too are its more superficial properties—malleability, conductivity, etc.—since these necessarily flow from the molecular structure of gold. Gold could not have the superficial properties of rubber and still be gold. Thus there are *de re* necessities at all levels of nature. The other paper I wrote was about the necessity of origin.Here I tried to explain Kripke's claim about essential parentage by generalizing the idea of origin. It isn't just parents and their children; it's grandparents, brothers and sisters, cousins; and the same holds for biological entities in general. It follows from this that I could not have existed unless a specific pair of apes existed in the past to propagate the evolutionary line that led eventually to me; indeed, without a specific set of primitive organisms in the dim distant evolutionary past I could never have existed. That is: There is no conceivable possible world in which I exist and yet the organisms that are actually in my ancestral line do not exist in that world. I cannot be detached from that line and placed in another line while retaining my identity. This means that even God could not create me without placing me in that line. He could create a person just *like* me—my exact twin—without bringing Joe and June into the story (and the whole biological tree that lies behind them), but he wouldn't thereby have produced me (clearly, I am not the exact same person as my twin, since we are two, not one). Just as God could not create a world in which the number two is odd, so he could not create a world in which I originate from something other than Joe and June; there is no world in which I was born to the parents of Elvis Presley, say. More precisely: My origin has to be in the particular sperm and egg that actually gave rise to me, not from the sperm and egg of people distinct from my actual parents. This gives new meaning to the idea of "family ties," and I daresay psychoanalysts would seize upon the necessity of origin as a vindication of the determining role of parents in fixing one's essence. In any case, parental bonds are as metaphysically strong as the evenness of the number two or the singleness of bachelors. Break them and you no longer have the same person.

I used the phrase "possible world" a couple of times just now. Possible worlds received a lot of attention in the mid-seventies, as people tried to figure what they meant by this useful phrase. It trips smoothly off the tongue and helpfully encapsulates our intuitive beliefs about what is necessary and what is contingent: If something is necessary then it holds in all possible worlds, whereas if it is merely contingent then it holds in only some possible worlds. But what is a possible world? Are there really such things? If so, where are they? And how do they compare with the actual world? David Lewis, now at Princeton, decided to go for broke on this question, maintaining that possible worlds—say, the world in which I became a drummer, not a philosopher—are just as real and concrete as the actual world. Lewis holds that other possible worlds contain "counterparts" of me, just as real as me, flesh-and-blood people, located in space and time, involving themselves in all sorts of things I would never get up to in the actual world. The actual world differs from these other worlds just insofar as I refer to it when I use the wold "actually," but not in its intrinsic ontological nature (those other worlds are also referred to by their inhabitants when they say "the actual world"). Possibility is like space, for Lewis: All the parts of space are equally real, including those that are not in my vicinity; it is just that we distinguish one part of space with the word "here" and give this a special prominence. To borrow a phrase of Kripke's, Lewis thinks of possible worlds as rather like distant lands that we cannot travel to, but are no less real than the place we inhabit.

As Lewis has often acknowledged, this theory of possibility is often met with "incredulous stares," as if people cannot believe he is saying what he is saying; but Lewis maintains that this picture gives the best overall account of the concepts of necessity and possibility. It is true that Lewis himself has a kind of otherworldly appearance, with a marked fixity of expression, a long straggly beard suspended from the underside of his chin, and a voice precise and robotic. Perhaps, to him, the space of possible worlds feels as real as the actual world all around him; I have heard it said that his car driving leaves a lot to be desired. In any case, Lewis's extreme "modal realism" has not made many converts, despite the brilliance of its main proponent. I wrote a long paper about it in the late seventies titled "Modal Reality," in which I denied the existence of possible worlds but insisted that necessity is still an objective trait of the universe. I was a realist about necessity but a skeptic about possible worlds. But I have to confess to a bit of a soft spot for possible worlds, as Lewis interprets them: It's strangely comforting to think I have all these milling counterparts in other possible worlds, living such different lives from me, some of them having a worse time than I (though others make me envious); and I like the sheer boldness of the Lewisian conception. But, as with the idea of heaven, though it would be nice if such alternative realities existed, I just cannot bring myself to believe that they do. There is only the actual world and its inhabitants (and it is quite big enough to contain everything there is).

The reader may be wondering at this point why he or she is not being treated to retrospective tenure anxiety, especially in view of the several close calls tremulously outlined in earlier chapters. Tenure: the ultimate torment of academic life ("tenure" and "torture" are remarkably similar words). The reason I am not belaboring the tenure trauma is that the tenure system in England at this time was not a very onerous one. So long as you did your job and wrote some publishable papers the head of department would typically recommend you for tenure automatically. Then, security till retirement. I had started publishing articles in good journals (*Analysis, Journal of Philosophy, Mind*) soon after arriving at UCL, so I wasn't in any real danger of failing to obtain tenure, and I did obtain it some two years after going there, without much fuss or fanfare. I was, it is true, youthfully anxious about establishing a good reputation, but tenure was not for me the nightmare ordeal it is for many young academics in America. It came easier than most things.

Not that everything went smoothly for me, however. A couple of years after coming to UCL I applied for three tutorial fellowships in Oxford. As I noted earlier, I had wanted to stay in Oxford after the B.Phil, and I still felt that Oxford was the center of philosophy in Britain; in addition London was expensive and inconvenient for someone on my very modest salary (I was still living in a single room in someone else's flat—though it was now a big room and I had a desk to write on). Since I had accumulated a fairly solid publishing record I had hopes that I would be appointed at one of the Oxford colleges. However, I was appointed at none of them, though I was interviewed at two. I felt at the time (and still do—though this is something one is not supposed to say) that I was passed over and that the people who were appointed over me were not necessarily the best candidates. Here I got my first sour taste of academic politics and the operations of cronyism and favoritism. Since then I have witnessed all too many cases of this, though not with me at the receiving end. But at the tender age of twenty-six I was naive, and innocently believed that people always wanted to appoint the best person for the job—and not, say, the person who would support their views and not rock the boat. I resigned myself to staying in London—hardly a dismal fate—and put the idea of returning to Oxford out of my mind. And I think, in retrospect, that this was probably the best thing for me; it allowed me to attain the kind of intellectual independence that is so necessary in philosophy. My thoughts did, however, start to turn toward the idea of moving to America, where I believed (rightly) that the competition would be fairer, and the philosophical world less dominated by a single institution. At the least I was keen to visit America and experience the philosophical atmosphere there, firsthand. I had hardly traveled out of England at all at this time, mainly for want of funds, and being consistently rejected by Oxford started me on the road to leaving the country in which I was born, though this was not to come about until quite a few years into the future.

An influential Oxford figure at this time was Michael Dummett, a philosopher of language and mathematics, who had spent most of his academic life at All Souls College, not teaching (All Souls is a research institution exclusively). He had published a large and influential book on Frege, the late-nineteenth-century German mathematician who had done so much to determine the shape of analytic philosophy. According to Dummett, the right way to do metaphysics was via philosophy of language: If you want to find out about the basic structure of reality, you should investigate how we speak about reality. This was part of the "linguistic turn" that characterized twentieth-century philosophy, but Dummett had bigger fish to fry than the average "linguistic philosopher." His ultimate interest was in the question of idealism versus realism—whether the world depends on our minds in some way or whether it is quite independent of our minds. This is a longstanding philosophical issue: The doctrine of idealism had been central in philosophy, following Berkeley, Hegel, Bradley, and others; what Dummett brought to this debate was a more sophisticated philosophy of language. The question became: Is the meaning of our sentences to be explained in terms of the evidence we have for asserting them, or is it a matter of truth conditions that have nothing intrinsically to do with evidence? This connects with idealism—the doctrine that the world is ultimately mental—because the evidence we have for our beliefs comes down to the sensory states that prompt us to believe what we believe. For example, if I assert, "There is a brown table in this room," I do so because of the evidence of my senses—that is, because it *seems* to me, visually, that there is a table over there; and this seeming is a state of my sensory system, a mental state I am in. So if the meaning of the sentence is to be explained in evidential terms, then the truth of the sentence will turn out to involve sensory states such as seeming to see a table. This is the so-called antirealist (or idealist) position. A realist, by contrast, will say that the sentence is about a table and a room that have their existence quite independent of any sensory states I might have or speech acts I might perform. For a realist, such sensory states or speech acts are a *consequence* of the existence of physical objects, as they causally act on our sense organs, but they are not what *constitute* physical objects. Realism holds that objects are intrinsically mind-independent causes of mental states that give us evidence of their existence.

Or consider statements about the past. I assert, "I went to high school in Blackpool," because of the memories I have of being there, but a realist about the past will sharply distinguish these memories from the fact I am stating, which could hold, whether or not I have such memories. The memories are my evidence for the claim about the past, but they are not what that claim is about. Hence the claim could in principle be true even if I have no memories of being at school there (say, if I have amnesia). Generally speaking, statements in the past tense can be true or false independent of anyone's having any present evidence that might enable them to decide the question. In philosophical jargon, statements about the past are subject to the law of "bivalence," which declares a range of statements to be determinately either true or false, whether or not we can discover which of these two "truth-values" they have. An antirealist, however, will say that my statement is really about my memories, since its meaning is given by the evidence I have for asserting it (this view is also called "verificationism"). One consequence of the antirealist position about the past is that, if there is no evidence either way with respect to the truth of some past-tense statement, then the statement cannot be said to *be* either true or false. Thus if I say, "There was a deer in my garden exactly ten years ago," this sentence will be neither true nor false if no one has any evidence

either way; yet, according to our ordinary way of thinking, either there was or there wasn't a deer in my garden
ten years ago, no matter what anyone can remember or
reconstruct. The past, for an antirealist, is really constructed from our memories of it and has no reality independent of these memories; while a realist takes the past
to consist in a set of facts that obtain whether or not we
have any such memories. The realist's point is not that we
can *know* about the past without having any memories of
it; that is obviously impossible. Rather, the realist holds
that the past *itself* transcends our present memories of it
and is not constituted by such memories. Reality is what it
is, the realist maintains, whether we can know about it or
not, whereas the antirealist thinks that reality is ultimately a product of our cognitive states. Just as the reality
of a fictional character in a novel is exhausted by whatever
the author puts into her story, so the world in general is a
creation of our human construction. And just as there is
no answer to the question of whether Hamlet had a mole
on his right shoulder blade—since Shakespeare neglected
to supply any details on the question—so the past will contain gaps where memory falters. Gaps in memory are gaps
in the past, for the antirealist, as gaps in Shakespeare's
description of Hamlet are gaps in *Hamlet*.

I totally disagreed with this antirealist picture of the
world. But in philosophy you have to be able to refute the

other guy's argument before you can claim to be on solid ground, so I had to examine Dummett's obscure and complex arguments in order to find out what was wrong with them. I spent perhaps too much time on this venture, in the end concluding that Dummett's arguments had a number of obvious flaws. The main problem was that he assumed from the start that understanding a sentence is a matter of being able to verify it. But a realist will question this right away: Why not suppose that understanding a sentence is a matter of knowing what would make it true, whether or not we can ever find out whether it is true? We have two sorts of knowledge about any particular state of affairs: what the state of affairs itself consists of, and what would count as evidence for believing the state of affairs to have occurred. Thus I know what it would be for there to be a table in the next room, since I have a conception of physical objects in objective space, and I also know what kind of evidence would make me assert this, namely the experience of seeing such a table. Why conflate these two pieces of knowledge? Only, it seems, to insist that the former knowledge be reducible to the latter; but that is just to presuppose antirealism (or verificationism) from the start.

The general point here is that it is wrong to confuse reality itself with our ways of knowing about it. Reality is one thing; our knowledge of it is another. The past is not the same as our memories of it; physical objects are not the same as the sensory states we have when we perceive them; other people's minds are not the same as the behavior we use to infer things about them; the future is not the same as the current indications of how it will turn out; elementary particles are not the same as the meter readings that signal their presence; and so on. To be sure, there are exceptions to this general rule; as already mentioned, fictional entities have no reality beyond the intentions of authors—they are invented, not discovered. That is why we call them *fictions*, and distinguish fiction from nonfiction in bookshops and libraries. Real detectives are not the same as fictional detectives—of course they're not.

And there are more contentious and interesting cases, such as color. Do objects have color independent of human vision? Colors certainly look to be out there, spread over the objects we see; they don't seem like items inside the perceiving mind. But can colors and experiences be completely independent? Let us consider a thought-experiment (the philosopher's equivalent of a regular empirical experiment): Suppose there are Martians who also have a sense of sight in which objects seem to have certain colors, and suppose they see the same range of colors as humans. However, suppose also that there is an odd inversion in Martian color perception: What we see as red they see as green, and vice versa. When I look at a rose and see it as red they see the same rose as green, but verdant grass looks scarlet to them. This is the way it is with all of them, except for a few color-blind Martians, and it has always been this way. Which of us sees the color of the rose as it really is? The question seems misguided; it can hardly be that we are wrong to see the rose as red and that the Martians are right, and it would be a pigheaded chauvinist who insisted that Martians can't see straight. Why can't both of us be right? If I like the taste of beetroot and you don't, it hardly makes sense to say that one of us is right about the tastiness of beetroot and the other is not. Or consider the taste of rotting meat to a vulture in contrast to how it would taste to a human. Color is like that; the color of an object is the color it appears to have to normal observers in normal conditions, but in the case of Martians what is normal for them isn't normal for us. What we need to acknowledge here is that roses that, to us, are red are green for Martians, with no error on either side. In other words, color is relative to a perceiver. We can make sense of perceptual error about color *within* a class of perceivers, as when jaundice makes everything look yellow to me; but we cannot make sense of such error *across* groups of perceivers. Color is in the eye of the beholder, basically. So we cannot properly dissociate the color of an object from the experiences the object produces in the mind; color is a mind-dependent aspect of reality. It is, as philosophers say, a disposition of objects to produce experiences in perceivers.

But that is not to say that *all* properties of objects are similarly mind-dependent. The shapes and sizes of objects are not mind-dependent in the same way at all: If Martians see as small and square objects that we correctly see as big and round, then they see them wrongly; we can't both be right about the shape and size of something if we see it quite differently. Shape and size are objective features of things, while color has a subjective nature. Thus philosophers traditionally distinguish between the objective, mind-independent "primary qualities" of things, and the subjective, mind-dependent "secondary qualities" of things.

Our ordinary perceptual point of view incorporates both subjective and objective dimensions, both colors and shapes (as well as the properties revealed by our other senses). But we also strive—especially in science— to represent to ourselves how the world is when shorn of subjective features; this is what physics aims to do and succeeds in doing. With physics it is all wavelengths and photons, not full-blooded colors. The way physics talks about the world is entirely objective, mentioning only properties that things have whether or not there are any minds to perceive these properties. (Physicists themselves may not always be entirely "objective," i.e., free of biases and prejudices—but that is a different question.) This raises a puzzling problem: Why don't we see the world only as it objectively is? *Could* we see the world thus? Might there be a form of vision that was limited only to objective features of things, no colors allowed? This would mean seeing the properties of things that physics posits, but not seeing those things as having color. But can we really make sense of *seeing* colorless objects? I don't mean seeing transparent objects or black, white, and gray objects, since these involve color too—either seen through the transparent object or because of the existence of achromatic colors. I mean seeing objects yet having no sort of color whatever in your visual field, not even a dull gray. This seems impossible to conceive of: There is no such thing as vision without *some* sort of color perception. So vision is essentially subjective; it *could* not give us only the objective truth about the world. The world as depicted in physics could not be a perceptual world.

I wrote a book all about this subject while teaching in London, entitled *The Subjective View* (published in 1982). What fascinated me about the subject was that the world as it is, in itself, independently of human minds, is not a world that the human mind could ever apprehend other than theoretically. We do really see physical objects and their properties, but we cannot expect to see them purely objectively, just as they are represented in physics; we are necessarily locked inside our subjective perceptual perspective. Nevertheless, human reason does enable us to get outside of our necessary perceptual subjectivity in order to form a representation of the world that is purely objective. We have concepts that contain no subjective taint, even though perception is irremediably subjective. This is a remarkable feature of human reason—its ability to transcend our subjective perceptual viewpoint and describe the world as it is, independent of that viewpoint. The human intellect works as a device of distancing from our subjective makeup. It is almost as though we have a subjective self *and* an objective self (Thomas Nagel, about whom more later, has written beautifully on this subjective/objective dichotomy in his book *The View from Nowhere*). I don't think other animals are capable of this kind of cognitive transcendence to absolute objectivity, being far more confined to their given perceptual point of view; we alone know how the world is constituted independent of our natural perspective on it. That is what science fundamentally is: a way of describing the world that abstracts away from human particularity and bias. The most obvious example of this "de-centering" is astronomy: We now see ourselves as occupying one small planet in a vast universe, no longer at the center of things, and subject to universal laws of nature—though this is certainly not the way things naturally appear to us.

When I think of these topics I recall my old friend Ian McFetridge, a very local presence on the London philosophical scene. He came to London, to teach at Birkbeck College at the same time as I, but came from Cambridge. He was a short, springy man with a small moustache, fiery brown eyes, and an ebullient manner. I started talking philosophy with him soon after we arrived, as we shared an interest in philosophy of language and logic. I appreciated his quick, darting intellect and his fine philosophical judgment. He was the kind of philosopher who saw one's point immediately and always had something to add to it, either critically or creatively. He could sometimes be a bit too animated, as if small explosions were being detonated in his head, but I liked his seriousness and sound philosophical sense. I also liked him as a person: He was humorous, generous, lively, compassionate, human. At the end of my teaching day I would often stroll over to Birkbeck to meet Ian, who taught mainly in the evening; if he wasn't in his office he was already in the pub. I would order my usual half pint of lager while Ian went through the pints of beer at an impressive pace. We would gossip and talk philosophy, sometimes with others in attendance. I would try out my latest idea on him, or he on me, and we always had an illuminating discussion. I valued his opinion of my work immensely. Then, after about five quick pints, he would hurriedly announce that he had to go and give a lecture. This never ceased to amaze me: I would start to lose my philosophical head somewhere through the third half of lager, while Ian would be perfectly coherent after his fifth pint, no doubt proceeding to give a scintillating lecture.

But Ian had problems. He worried obsessively about his work, lacerating himself for lack of productivity, and everything he wrote seemed to emerge from a deep well of anguish (what he did write was excellent). This made me uncomfortable, because I was happily churning the stuff out. Still, he was always generous about my work. He smoked and drank far too much, despite his ability to hold his booze (or perhaps because of it). He also had difficulty reconciling himself to his homosexuality, though he was not slow to act on it. All this was combined with a manic-depressive temperament that grew more pronounced over the years I knew him. He committed suicide in 1987, apparently during a bout of depression and in a drunken state. I had not seen much of him for a year or so beforehand, having moved away from London by then, but I heard from friends that he was having psychological trouble. I miss him as a philosophical friend and as an exceptional human being, and I remember the happy times we spent together as young men arguing about philosophy in the pub. And of course I wish I could have done something to prevent the terrible tragedy of his early death. There is now a seminar room in the Birkbeck College philosophy department that is named after him, a testament to the regard in which he was held.

My years in London also saw a resurgence in my interest in the mind. I had put this aside since leaving psychology some years before, and I had a kind of phobia that I would be forced to return to psychology (I would sometimes dream that I was still, discontentedly, a psychologist). For years I regarded my psychological background as if it were a shady, disreputable past that I needed to keep quiet about if I was to be taken seriously as a philosopher. But a number of factors conspired to draw me back to the mind. First, some of the philosophers of language I was most interested in had also written significantly on the philosophy of mind—Kripke, Davidson, Quine, Putnam, and others. Since their work was interconnected, I couldn't very well avoid the portions written on the philosophy of mind. Second, there was a renewed interest in the philosophy of mind within philosophy itself, as it became apparent that many of the issues in philosophy of language were really issues in philosophy of mind, particularly those pertaining to reference. Third, I couldn't stop myself having ideas about the mind—which had, after all, been my main object of study for several years. Fourth, after enough years had passed I ceased to be afraid that I was nothing but a psychologist in disguise; by this time I had become a philosopher through and through. I could now with confidence approach the mind philosophically.

I had two main interests in this subject, which I shall now briefly describe (I shall return to them more fully later). I wrote in chapter 3 about proper names and reference: The question was how a name manages to refer to an object. It started to become apparent in the seventies that this question was really just an indirect expression of another more fundamental question: How do thoughts manage to be about things? If I say, "Plato wrote *The Republic,*" I have a thought to the effect that Plato wrote *The Republic*, and my thought has "intentionality"—that is, my thought is about Plato. We can then ask: What *makes* my thought be about Plato? Not surprisingly the same kinds of theories suggest themselves here as for the reference of names: Is it that I have in my mind a bunch of descriptions that Plato uniquely satisfies, or is it that my thought is a link in a long causal chain leading back to Plato himself? Asking these questions led, among other things, to the idea of a "language of thought" that encodes the process of thinking: In my head is a word for Plato (not necessarily the word "Plato") that refers to Plato by some means or other. The idea that the mechanism of reference of such mental symbols is a kind of causal connection between the object and the symbol began to gain dominance. Thus it was held that the appropriate word in the language of thought refered to Plato in virtue of the fact that Plato (the man himself) was the dominant causal source of occurrences of that word in the formation of beliefs. In simpler terms: What you think about is a matter of which objects cause what symbols to pop into your head. Thinking is a kind of silent speech in which symbols are causally hooked up to external objects. At the same time, the related question of what constitutes the content of a thought came to the forefront of discussion: When you think, you always think *that* something—for example, that London is the capital of England. But what gives a thought that content? And what kind of thing is a thought-content? (I shall discuss this more fully in the next chapter; I mention it now to show how my interest in language was transformed into an interest in mind.)

The second big issue that occupied my interest was the notoriously tricky mind-body problem. This is the question of the relationship between our mental states and processes, on the one hand, and the states and processes in our brain, on the other. Suppose I see a yellow bird and accordingly have an experience of bright yellow. At the same time the neurons in a certain part of my brain engage in a series of complicated physical processes involving electrical and chemical interactions. How exactly are the experience and the neural processing connected? Dualists say that they are quite distinct from each other, merely running in parallel; you could even have one without the other, in principle. Monists say that they are one and the same thing viewed from two different perspectives; you therefore could not have one without the other, since you can't have a thing without itself. This latter view is called the "identity theory": It holds that mental and physical states are literally identical, just as Bill Clinton and the president of the United States in 1999 are identical. The identity theory implies that facts about the mind are *reducible* to facts about the brain, so that there is nothing more to the mind than neurons in action. And there are other theories too: Functionalism, for example, says that mental states are really a matter of the *role* of brain states in the overall economy of our psychological life.

The view I was attracted to in those days was that every mental event or state could be identified with a brain state of some sort, but that it is not possible to reduce mental concepts to physical concepts. This view is often called "nonreductive materialism," or "anomalous monism," and it was championed particularly by Donald Davidson. It seeks to cleave to a basically materialist worldview without making the mind nothing but the brain. But this whole subject underwent a radical transformation in my mind some years later, which I shall come to in due course. Back in the seventies I was struggling with a difficult problem and not finding anything that really satisfied me. Something had to give, but in those days I had no clear idea what.

My first philosophy book circled gingerly around the mind-body problem. I had published a few papers on the philosophy of mind, and Oxford University Press asked me if I would write an introduction to the philosophy of mind for them. I agreed and gave an introductory lecture course on the topic as a preparation for writing the book. I wrote the book in six weeks during the term, followed by another month for polishing it. I called it *The Character of Mind*, echoing Gilbert Ryle's *The Concept of Mind* (a very different book from my own), and it came out in 1982, to good reviews, when I was thirty-two; it is still in print, now in a revised edition. It is a plainly written book, pithy, with a minimum of jargon, but quite tough-going in places. I am told that it has been useful to many generations of students, and I am glad that I wrote it, though at the time it seemed like a diversion from my research interests. Heretofore I had not thought of myself as a book writer, preferring to write papers, the shorter the better. Writing *The Character of Mind* gave me a taste for the longer form and made me realize that writing a book was not the endless labor it appeared to me to be. It is true that writing philosophy is never easy, and that a philosophy book is a demanding undertaking, but at least I discovered that I could sustain myself over the longer haul. I still have a taste for brevity, however, and try to keep my books short and sweet. There is nothing more irritating to me than a lengthy philosophy treatise in which the author never quite comes to the point, or labors the obvious. I always want to say: Get on with it!

My overall recollection of my years in London is of a decade of effort. I worked hard and took minimal vacations. As soon as the term ended I would go straight to my own writing. I read a great deal of philosophy. The tutorial system was an exacting taskmaster. I needed all the mental energy I could muster. But it was a time of intellectual excitement for me, as I began to make my way in the world of professional philosophy, making a number of good friends along the road. I had no clear idea what I might amount to in philosophy, but I was going to give it my best shot. Looking back, I marvel at my dedication and willpower. I had fought hard to become a philosopher and I wasn't going to waste the opportunity. I enjoyed seeing my name in print—what writer doesn't?—and was not above getting praise for my work. I was, however, still rather unformed and philosophically confined, too much a product of my background in Oxford. It was to take an entire shift of continent before that would change.

CHAPTER FIVE

Belief, Desire, and Wittgenstein

UNTIL 1979 MY PROFESSIONAL LIFE REVOLVED AROUND OXFORD and London, two places not far apart (you go to either of them to escape the other). But my mind was constantly on America; the philosophers I found the most interesting were American, and I preferred to publish my work in American journals, if I could. At this time Britain was becoming decidedly subsidiary to the USA in philosophical heft and influence, despite a British preeminence for most of the twentieth century. America was where the action was, philosophically. American philosophy was so much *bigger*, for one thing, and much better funded. But it was also the center of philosophical innovation, with Harvard, Princeton, Berkeley, UCLA, and elsewhere producing the most influential work. I therefore acquiesced with enthusiasm to Richard Wollheim's suggestion that I take time off from UCL to visit an American university. He had a contact at UCLA and made inquiries. After looking at my vitae and published work they invited me to visit for two semesters (six months), beginning January 1980, as a visiting assistant professor. This seemed like quite a coup to me, a mark of arrival: I was exportable.

There was a snag, however—I couldn't drive. There had been no point in learning, since I would hardly have been able to afford a car in London on my meager starting salary anyway. But driving, I was assured, is a sine qua non in Los Angeles, so I set about learning to drive in short order. I passed my driving test only a few days before leaving for L.A. and I really don't know how I would have managed if I had failed. Still, I was a novice driver, accustomed to small stick-shift cars, and imbued with the stubborn conviction that cars naturally travel on the left-hand side of the road. My road prejudices would have to change, as would my philosophical prejudices.

The English winter was exceptionally cold and wet that year, when I boarded the fourteen-hour nonstop flight for L.A. I had paid a visit to America once before—to New York—in the winter, staying mainly in Brooklyn. But California was quite new to me. I knew it mainly from the movies. I was blearily picked up at the airport by Tyler Burge, a young philosopher from UCLA whom I had met a year or so earlier in London. He drove me to a room on campus in which I was to stay for the weekend before I found an apartment to rent, and I went straight to sleep, exhausted from the trip. The next morning, waking early, I was astonished by what greeted me: palm trees (I had never seen one before), hummingbirds (I thought they existed only in zoos and jungles), and warmth, *warmth* (in January!). It really was like that scene from the *Wizard of Oz*—always one of my favorite movies—when Dorothy wakes up in Technicolor Munchkin Land, far from her black-and-white childhood. After the drab, gray, cold English winter it amazed me to find L.A. almost tropical and to be taking it completely for granted. I decided to take a walk to the beach, not realizing that the L.A. streets were not made for walking. After a couple of hot hours, with the ocean not getting perceptibly closer, I had to catch a bus back to campus, for which I did not have change; I was going to have to learn a thing or two out here.

I ended up living in a spare room in the apartment of the philosophy department's librarian, near the Mormon church in Westwood, not far from the UCLA campus. As it happened, a young Finnish philosopher I had met in England a few months earlier, Esa Saarinen, was also visiting at UCLA, and I soon fell in with him (we are still close friends to this day). He already owned a big American car and was a far more confident driver than I; with his help I bought a 1963 Chevy Impala, turquoise and white, for

$400. It was huge, and had no power steering, and the horn would automatically sound whenever you took a right turn. Still, it was quite an experience for me in those days to sit in it in the L.A. heat and cruise down Sunset Boulevard with the radio on. Fortunately, I had no accidents, though I did get a ticket for mistakenly taking an illegal left turn (I tried to explain to the officer that I was a British philosophy professor new to L.A., but I'm afraid he took a hard line). I also once started driving on the left-hand side of the road at night, and only realized my mistake when I saw headlights puzzlingly approaching me head-on. (My Finnish friend, Esa, took to calling me "The Ace" in ironic recognition of my not-always-stellar practical abilities, and still calls me by that nickname.) I asked a colleague in the philosophy department to come out with me in the car to give me a few pointers, and as we stopped at a light two young guys drew up beside us and said, "Say, do you dudes race?" Instead of explaining that he was just giving me a driving lesson, my companion, Warren Quinn, smoothly replied, "I think you've got the wrong guys." Before very long I was tooling along by myself, enjoying the breeze, singing along to the radio (I remember particularly "My Best Friend's Girl" by The Cars), trying not to look too much like a transplanted British philosophy professor. I found, though, that police cars would often tail me for a few minutes, simply because of the kind of car I was driving—and, I suppose, because of my youth and long hair. On one occasion I was even stopped for suspected drunk driving (the car did have a tendency to veer to the left for no reason) and given a Breathalyzer test; I tested negative, so the officer admitted his mistake—but it was an unpleasant experience, nonetheless.

But it wasn't all driving in L.A., though it sometimes seemed that way; I had philosophy to do. I taught a graduate seminar on realism and an undergraduate class on theory of knowledge. But the main topic of interest in those days was belief and desire: It seemed as if everyone in L.A. was talking about belief and desire (and I don't just mean belief in yourself and the desire to make it). Philosophers like to dignify these two pillars of the mental life with the label "propositional attitudes," because they involve taking an attitude toward a proposition. Take the proposition that it is sunny today: I can believe that this is true, desire that it be true, hope that it will be true, feel disappointed that it is not true, and so on. Propositions specify states of affairs, and beliefs and desires are attitudes concerned with states of affairs. The mind is very largely constituted by being the place where propositional attitudes do their thing: believing, reasoning, desiring, hoping, regretting—this is what mental activity is all about. You act as you do because of the beliefs and desires and other propositional attitudes you have: You desire to drink a beer, you believe that there is beer in the fridge, so you head to the fridge. Propositional attitudes constitute the reasons for action, the forces that shape human behavior. They are what make us tick. Not surprisingly, then, they have been of great interest to philosophers of mind. Here are some of the ways philosophers have demonstrated their interest.

As Frege long ago pointed out, propositional attitudes show that there is more to the meaning of a word than simply what it refers to. You might think that a proper name, like "Marilyn Monroe," simply means what it stands for—a certain famous bleached-blond actress. Isn't Marilyn Monroe more than enough to determine the meaning of her name? But there is a simple proof that this equation of meaning and reference has to be wrong. Consider the name "Norma Jean Baker," Ms. Monroe's real-life given name; it too refers to Marilyn Monroe, since Marilyn Monroe *is* Norma Jean Baker. One woman, two names. But suppose someone—call him Jack—believes that Marilyn Monroe is a great comic actress while holding no opinion about the acting skills of a person called Norma Jean Baker. If you ask Jack what kind of actress he thinks Norma Jean Baker is, he will reply, "Don't know, never heard of her." Or if Jack went to school with Norma Jean and is not aware that she became a famous actress, he may say, "Oh Norma Jean, she couldn't act her way out of a paper bag." Clearly, Jack believes that Marilyn Monroe is a great actress, but he doesn't believe that Norma Jean Baker is—he may actively disbelieve this. He assents to one proposition and dissents from the other. Yet the two names refer to the same person. So there must be more to the meaning of a name than simply its reference. On the basis of these kinds of considerations, Frege suggests that names have a "sense" as well as a reference, whereby the sense is constituted by the specific way a person thinks about the reference of the name. When Jack uses the name "Marilyn Monroe" he may think of her as the actress he saw in *Some Like It Hot*, but this way of thinking is not something he associates with the name "Norma Jean Baker"—here he may be thinking of her as that rather plain brunette he went to high school with in a small town in the Midwest. Same reference, different sense. To take Frege's own famous example: I may think of Venus in two ways, either as the planet I see first in the morning or as the planet I see first in the evening, these being two properties of Venus. If I use two names for Venus—say "Hesperus" and "Phosphorous"—that are associated with these two "modes of presentation" of the identical planet, then these names will differ in their sense, though coinciding in their reference. Reference does not determine sense.

This distinction between sense and reference enables

us to keep track of the difference between thought and reality: Reality is the reference, the thing in the world we are talking about, a person or a planet in our examples; thought corresponds to the sense we bring to our terms, the way in which we conceptualize the things we refer to. If we confuse sense and reference, we will be liable to think that objects are somehow mental or that concepts are somehow physical, but we always have to distinguish carefully between the external thing we are thinking about and the mental act of thinking about it in a certain way. Sense is rather like visual perspective: We may both see the same external object, but from different angles, so that it presents two different visual appearances to us; there is one object and two ways of apprehending it. There is reality, on the one hand, and then there are the many perspectives from which it may be viewed, on the other. This fundamental dichotomy manifests itself in the distinction between the sense and reference of our words. There is *what* we talk about (reference) and *how* we talk about it (sense).

What I have just described is the orthodox view of sense and reference. But there were two philosophers at UCLA, David Kaplan and Keith Donnellan, who questioned this orthodoxy. I would occasionally catch a glimpse of Professor Donnellan sitting silently in his office, seemingly becalmed, wearing white pants and dark glasses, only rarely volunteering a terse remark. One famous example concerns a man drinking water from a martini glass at a party: If you use the definite description "the man drinking a martini" while staring right at him, do you refer to him or not? He doesn't literally fit your description, since his glass has no martini in it, yet there seems to be some sense in which he was the target of your speech act. If so, sense and reference can become decoupled, and reference seems to trump sense, to bypass it. Professor Kaplan, in contrast to Donnellan, was never seated, seldom still, and notoriously loquacious—witty, clever, unstoppable, with a passion for proper names and the demonstrative "that." The license plate on his Mercedes convertible read "DTHAT," after the logical operator he had introduced some years earlier (it means roughly "the thing that is *actually* F," as in "the actual inventor of rock 'n' roll").

I got to know both Donnellan and Kaplan quite well, though Kaplan certainly said a few thousand times more words to me than Donnellan ever did. We shared a passion for the topic of reference—how our words contrive to pick out things in the external world. Both these philosophers were strongly opposed to the "description theory" of proper names, which I discussed in chapter 2 in connection with Saul Kripke. They were inclined to accord the reference of a name far more of a role in determining its meaning than the description theory could allow. They opposed the idea that the reference of a name is external to its meaning, with sense doing all the work, so that a name could have a perfectly good meaning, though it had no reference. Suppose I introduce an "empty name," say "Herbert," which I stipulate to refer, falsely, to the person who stole my watch (I lost it and don't want to admit to my carelessness). I go around saying things like "Herbert is a scoundrel": Does the name "Herbert" really have a meaning? The name seems semantically defective, a mere mark or sound. So, simply stipulating a synonymy between a name and a description seems insufficient to give the name a solid meaning. The name needs a real bearer if it is to function as a genuine name. Does this bearer displace Fregean sense, or is it best viewed as something additional to it? The problem was always what to say about names that have the same reference but differ in meaning. But there did seem something right in the idea that Frege had ejected the reference from the meaning in too radical a manner, and the description theory of names did have the severe problems Kripke and others had pointed out. How can the reference of our words be brought into closer relationship with what they mean without losing the distinction between sense and reference? That was the nature of the problem, as I conceived it.

Kaplan's proposal came to be called the "direct reference" theory. Instead of supposing that a name makes

reference via a description, Kaplan suggests that the reference is *direct*: There is no mediating concept, just the
reference itself. Thus the bearer of the name—a real-
world object—comes to be a component of the proposition expressed by the name as it combines with other
expressions. For example, the proposition expressed by
"Marilyn Monroe was a great actress" consists of Marilyn
Monroe herself, that flesh-and-blood woman, along with
the attribute of being a great actress. The object finds its
way into the proposition and hangs out there, displacing
any description the speaker might associate with the
name. This means that when you have a belief in that
proposition, Marilyn herself is a component of your
thoughts: She sits there in the proposition that you
believe to be true—herself, not the descriptions you associate with her name. Just by thinking of her you can bring
her bodily into your mental landscape. Such are the ben-
efits of "direct reference."

But, as I noted a paragraph back, there is the question,
How does this square with the fact that two names can
have the same reference and different senses? Granted,
the reference can insert itself into the meaning, but is it
then the *sole* component of meaning? Must we say that
"Marilyn Monroe" and "Norma Jean Baker" have the
same meaning after all, despite the fact that they cannot
be freely substituted one for the other in contexts of

propositional attitudes? The answer to this, which I developed (with a little help from my friends), is that there are really *two* aspects to the meaning of a name or other referring expression: its reference, and the way in which we conceptualize this reference. The orthodox Fregean view says that the meaning of a name is *just* the way we think of the reference, so that the object we refer to is external to its meaning, a mere semantic appendage. The Kaplanian view is that the meaning of the name is *just* its bearer, so that the way we think of its reference drops out. The view I favor is that the meaning is *both* the reference *and* its mode of presentation. I called this the "dual component" theory of meaning.

The theory is most easily illustrated with expressions like "I," "now," and "here"—expressions that are usually called indexicals. Suppose I say "I am hungry": I thereby use "I" to refer to myself, and what I say is true if and only if Colin McGinn is hungry at the time of utterance. Now suppose you say "I am hungry": This will be true if and only if *you* are hungry—I am irrelevant so far as your statement is concerned. There are two different references on two occasions, depending upon who utters the sentence, but obviously the word has something in common on these two occasions—it's just the identical word "I" that follows the semantic rule that it refers to whoever is uttering it. If all we had to play with were the two references, then we could find nothing in common between the two uses of "I"; but clearly there is more to the meaning of "I" than that. The best view, then, is that the meaning of "I," when I use it, consists of both its reference—namely me—*and* the common feature that sticks with the word when you use it to refer to yourself. The proposition is a kind of combination of these two elements, a bit of both. Thus we can acknowledge that the reference enters the proposition expressed, while not falling foul of Frege's problem about nonsynonymous words that have the same reference. It is not that "I," when I use it, means the same thing as "Colin McGinn" because these words have identical reference; rather, the meanings differ, because "I" has more to its meaning than simply its reference on an occasion. So we can agree with Kaplan that objects hop into propositions and hence become constituents of our thoughts, but we also hold that there is more material inside propositions than those objects—they're also the way we think about objects. Our beliefs and desires are immediately engaged with the world of objects, but these beliefs and desires also have their own inner lining, their own conceptual padding. We therefore get the best of both worlds; and since getting the best of two worlds is better than getting the best of only one, I liked this dual component picture of meaning—even though some people found it too messy and complicated. (Was this a typically Californian desire on my part to "have it all"?)

There is also the question of how our propositional attitudes relate to the language we speak. The main theorist here was H. P. Grice, an Oxford philosopher transplanted to Berkeley, who had a reputation for being the cleverest philosopher you could ever meet (he died in the early nineties). I never met him or even set eyes on him, though that was said to be an experience in itself: only one odd, discolored tooth in his mouth, an enormous potbelly, and a reputation of being the worst-dressed man on the continent—though widely agreed to be utterly charming. From his writing I knew him to be a philosopher of formidable intelligence, an original thinker, and an elegant (if over-elaborate) writer; I wish I had met him, but it was not to be. Grice was interested in what is involved in meaning something by your words when you communicate. I may use the words "the weather is awful today," thereby meaning that the weather is awful today, the conventional meaning of those words; but I might have used the same words to mean something very different—as when I use them as a code between you and me to mean "the person talking to us is a terrible bore." Grice called this "speaker meaning" and distinguished it from the literal meaning of the words I utter (which is that the weather is awful today, on both occasions of utterance). His question, then, was, "What is it for a speaker to mean something by his or her words?"

The obvious answer is that this involves a kind of communicative intention. If I say, "the weather is awful today," meaning that the man talking to us is a bore, I intend to produce in you a certain belief—namely that I think the man in question is boring. So, is speaker meaning just a matter of having such an intention to produce a belief in my audience? Surprisingly, the answer is No, as Grice ingeniously pointed out. Suppose I want you to believe that a certain enemy of mine has been rifling through your stuff, so that you will be angry at him. I go into your room, rifle through your stuff, and intentionally drop a glove that closely resembles the gloves habitually worn by my enemy. I know that you will find the glove, recall the gloves my enemy always wears, and assume that it was he who rifled through your stuff. I thus intend to produce a certain belief in you, to the effect that my enemy rifled through your stuff, by dropping the glove in your room. But do I thereby *mean,* by dropping the glove, that my enemy rifled through your stuff? Apparently not: I want you to form the belief in question, but not as a result of my meaning anything by my act. Why? What is the difference between this case and a normal case of communication? Grice gave the following answer: What is missing is that I have a *further* intention, namely that you recognize that I have the first intention. In the case described I certainly *don't* intend you to recognize my intention to deceitfully make you think that my enemy rifled through your stuff; on the contrary, I want to keep this intention hidden. My intention to produce the belief was not intentionally overt, as it is in the standard case in which I say something and mean something by it. Contrast the case I just described with a case in which we have a prearranged code to the effect that when I drop a glove in your presence you are to take it to mean that I could use a drink. Here, when I drop the glove I *do* intend for you to recognize my intention to produce in you the belief that I could use a drink. So what makes an act a case of communicative meaning is that there is this second-order intention that my first-order intention should be recognized. In other words, my intentions should be *transparent* to you. That is the essence of meaning something to somebody by something.

The importance of this analysis of speaker meaning, aside from its intrinsic interest, is that it offers the prospect of analyzing the whole phenomenon of linguistic meaning in terms of propositional attitudes. If we can analyze the notion of speaker meaning in terms of intentions and beliefs, maybe we can analyze sentence meaning in the same terms, once we add a provision to the effect that it is a *convention* to mean a sentence in a certain way. It is not a convention to use "the weather is awful today" to mean that a person is a bore, though it is possible to use it so; but it is a convention to use that sentence to mean that the weather is awful today (and other languages choose different linguistic strings to be governed by the same convention). So maybe linguistic meaning is speaker meaning *plus* convention: Words mean what they do because speakers have conventions to use them with certain sorts of second-order intentions. Semantics thus turns into a department of the philosophy of mind, because meaning is a matter of underlying propositional attitudes. Meanings are not, then, ethereal abstract entities floating around in Platonic heaven; rather, they are projections from the humdrum business of belief, desire, and intention. Indeed, if we could just analyze intention in terms of beliefs and desires—as, say, the belief that I will do something or the strong desire I have for something—then we could get the whole thing down to belief and desire: a huge simplification. And then, and then—what if we could go on to reduce belief and desire to brain states? Then we would have a total explanation of meaning and language in physical terms. Philosophers of a materialist persuasion sighed with yearning at the prospect of such a reduction. Much work was done trying to make the project work, and the mind-body problem assumed an even greater importance in light of this approach to meaning. The main problem for the approach was that we had to assume beliefs and desires as foundational, and the question of *their* intentionality still loomed large.

Let me also mention another idea of Grice's that deserves to become common currency. This is a distinction between two sorts of implication—the sort that words have and the sort that people convey. If I am asked for my opinion of another philosopher's abilities, and I don't think much of them, I might reply, pseudo-tactfully, "Well, he certainly has a clear speaking voice." Here I obviously imply that I have a low opinion of his philosophical talents, though I utter no words that say this. The words I utter simply express the proposition that he speaks clearly, which I may well sincerely believe; but I know that by *not* explicitly expressing a belief about his philosophical abilities I will lead my audience to believe that I think poorly of them. My words do not in themselves logically imply that the person has questionable philosophical talents, but *I* certainly imply this by using these words as I do. Grice called this "conversational implicature," and distinguished it from ordinary logical implication, urging that we not confuse the two. His point was that we can *say* what is perfectly true in a conversational context and at the same time conversationally *imply* something quite false, and the latter does not contradict the former. Suppose I actually think very well of someone's philosophical abilities, but I dishonestly wish to downplay this opinion, perhaps because I am a rival of this person's. I am asked what I think of him and I reply with my sentence about his excellent speaking voice. Here I *say* nothing false, but I *convey* something that is quite false—that I think he isn't very bright. Dishonest speakers, politicians prominent among them, use this technique all the time: Say something you can't be faulted on, but conversationally imply something that you want to put across. That way if anyone contradicts what you intentionally put across you can always craftily reply "but I never *said* that." Here is an example.

Interviewer: "What, Senator, do you think of the allegations about your opponent's pot smoking?"

Senator, knowing full well that the allegations are false: "I think anyone who isn't hard on drug crime has a dubious past."

Later interviewer: "We have it on record that in 1988 you officially rejected the allegation that your opponent had smoked pot, and yet yesterday you claimed that the allegation is true."

Slightly squirming, but blustering, senator: "I never *said* my opponent ever smoked pot, I simply said that those who are soft on drug crime very often have a dubious past, and let me tell you that my opponent's policy on repeat drug offenders is what is responsible for the addiction of our children—and that is only a part of the problem. . . . taxes . . . welfare . . . defense . . . the flag."

Armed with Grice's distinction, the interviewer could have shot back with "I know you didn't utter a sentence that logically implies that your opponent smoked pot, but you certainly conversationally implied it, and that is what I want your response on now." There are two ways you can tell lies: by uttering outright falsehoods or by saying what is true and conversationally implying what you know to be false. The latter is no better morally than the former, but it gives the liar an escape clause, should the need arise. But now we are straying into moral philosophy, and away from philosophy of language.

Those six months in Los Angeles changed me philosophically. As a result of contact with the philosophers there I shed my lingering attachment to the Oxford philosophical scene. I became far readier to think outside of the accepted framework, and far more critical of the ideas that were then current in Oxford. It has always had a reputation for nurturing a hothouse atmosphere and an inward-looking obsession with its own workings. I saw this clearly now, and I didn't want my own work hampered by the style and assumptions that prevailed at Oxford. The intellectual air in L.A. was freer, whatever may be said of the actual physical air. I was highly impressed by many of the philosophers I met there, and I felt that they were more receptive to me than the Oxford establishment, many of whom seemed to regard me as something of a pest for not toeing the party line. When I returned to London in 1980 I felt both rejuvenated and discontented. I would have dreams about L.A. and my Chevy Impala, and I missed my American friends. England weighed heavily on me.

But it wasn't to be very long before I was back in California. A year later the University of Southern California offered me a job. I had several philosophical friends there—Hartry Field, Brian Loar, Stephen Schiffer—and I was very tempted by the offer, but it seemed like an awfully big step, and I wasn't sure that L.A. suited me as a city: too big, bland, and modern. I asked if I could come for a visit at USC so that I could decide whether I wanted to make the Big Move, and in 1982 I was back. This visit didn't go quite as well as the first time. I drove a truly awful cheap Chevy Nova; it rained nearly all the time; and I had my car battery stolen in the parking lot at USC, leaving me stranded in a risky part of the city at midnight. USC was too far from Brentwood, where I lived; the freeway traffic could be terrible; I skidded across Sunset Boulevard one day when the surface was slippery from oil and rain. The centerlessness of L.A. began to get to me, and the loneliness. The philosopher lost in La-la land.

It was during this period that I developed an obsession with video games, spending many a rainy hour in the amusement arcade in Westwood Village. I used to play a lot of pinball as a teenager, on the piers of Blackpool, and was pretty handy at it, so video games were a natural next step. It was Hartry Field, a high-powered philosopher of mathematics—whose shirt seemed incapable of staying tucked into his pants for longer than five minutes— who introduced me to my first video love, Ms. Pacman. One day in a bar near Venice beach Hartry asked me if I wanted to play a game; I said I'd never tried, preferring pinball, but I'd have a go. That was the beginning. I noticed an itch to play again the next day and called Hartry. Soon we were meeting almost every evening for furtive sessions with the chirping, gulping melon, playing for hours at a time. Often it would be painful to drive home afterward because my right arm was so strained from slamming the lever around. Then I started going solitary, feeding my addiction. My obsession with Ms. Pacman eventually shifted to Galaga, a game of shooting, not gulping. Even now, nearly twenty years later, I can still see and hear the icons as they dove from the top of the screen, and I can feel my shooting fingers start to twitch, the adrenaline rushing. I would park my battered Chevy near Wilshire Boulevard and take the ten-minute walk to the UCLA library through Westwood Village, but invariably I would be drawn to the amusement arcade for a "quick game." Two hours later I would blink into the L.A. sunlight, bleary, frazzled, twenty bucks poorer—but onto stage thirteen at last! Much later I moved on to

Defender, a game so demanding, so all-consuming, that I began to understand all those stories about teenagers hopelessly lost to video games. I became an arcade addict, a machine machine. But I will fight the temptation to dilate further upon this ludic phase of my life, lest the reader suspect I am still not over it (I haven't played a game in years, honestly).

The obsession with video games went along with another nerve-fraying obsession at this time: Wittgenstein. Both took abnormal amounts of concentration, enormous persistence, and a slightly masochistic taste for frustration. Wittgenstein is one of the most famous names in twentieth-century philosophy, both inside and outside the subject. This is not simply because of his philosophical ideas, which are dense and difficult to follow; it is his personality that intrigues people. He lived a philosophically pure life, giving away his large family inheritance, never owning a home, forever disappearing from Cambridge for solitary contemplative winters to coldest Norway in a self-made shack. He took the life of the mind with utmost seriousness. He was also morally extremely intense and fierce. He has an enormous following outside academic philosophy (and a large one within it). This is partly because his writing is oracular, almost poetic, and lends itself to multiple interpretations; but it is also because of his proud unworldliness, his intellectual dedication. He perfectly fits the model of the lonely, tormented genius. His ideas, however, are mainly confined to technical analytic philosophy of language, mind, and mathematics. Above all, Wittgenstein is uncompromising—a quality both admirable and infuriating.

Kripke had published a long paper on Wittgenstein that was causing a bit of a stir; this was undoubtedly helped by the perception that Kripke was the nearest thing to Wittgenstein in contemporary philosophy. I thought it would be interesting to do a seminar on Kripke's paper, with my friend at UCL Malcolm Budd. I hadn't really studied Wittgenstein very carefully before this, finding his work too unsystematic and impenetrable. But Kripke was always clear, and he seemed to have an interpretation of Wittgenstein's *Philosophical Investigations* that made coherent sense of it. Since Kripke's *Naming and Necessity* had revolutionized much of philosophy, I expected his new work on Wittgenstein to do the same. I studied his paper enthusiastically and did indeed find it lucid, systematic, challenging, and original—Kripke had done it again! With Kripke's interpretation of Wittgenstein in mind I could now read the *Investigations* and expect everything to fall into place. On the weekend before the seminar was due to start I set about reading Wittgenstein with Kripke's paper to hand, fully expecting a total vindication of Kripke's interpretation. But as I read the *Investi-*

gations—grappling from one section to the next, hanging on tight—doubts sprang up, and by the end of the day I started to suspect that Kripke had it seriously wrong. I just couldn't see that Wittgenstein was saying what Kripke said he was. But I was no expert; perhaps Kripke's interpretation shone through more clearly in other works by Wittgenstein which I hadn't yet studied. I went to bed on Sunday night feeling thoroughly perplexed.

The next day I saw Malcolm in the department: He had written his Ph.D dissertation from Cambridge on Wittgenstein and I respected his opinion. I remarked, tentatively, that I was having a difficult time seeing that Kripke was interpreting Wittgenstein correctly, and stood ready to cite the passages that seemed to me to contradict his interpretation. Calmly, Malcolm replied: "Kripke isn't right at all—Wittgenstein is saying something totally different." We then compared notes and found that we basically agreed on what Wittgenstein was saying and where Kripke had gone wrong, though my knowledge of the Wittgenstein corpus was markedly inferior to Malcolm's. As the term progressed I studied Wittgenstein's work more thoroughly and found my initial opinion corroborated. By the time the term was finished I had a worked-out interpretation of Wittgenstein and detailed criticisms of Kripke. Since most people seemed to be agreeing with Kripke, I decided to write a paper or two on the subject, which would require me to immerse myself still further in Wittgenstein's challenging writings. And this immersion was my condition at USC, where I also taught a graduate seminar on Kripke and Wittgenstein. Rain, mud, video games, Wittgenstein—that was my L.A. for those four peculiar months.

Let me try to give a concise account of the two contrasting views of Wittgenstein's message in the *Investigations*. The question concerns what it is to follow a rule, say the rule that if you put "+" between two numerals the result is the *sum* of the two numbers you are specifying. The rule for "+" is that this sign denotes addition, and anyone who understands "+" knows that it stands for the addition function and not some other mathematical function. But what is this understanding of "+"? What *is* it to grasp the rule for adding one number to another? Following Wittgenstein, Kripke makes a very important point about this, namely that it cannot just be a matter of applying the rule in a finite number of cases: It must amount to more than giving the sum of any pair of numbers for any finite series of arithmetical questions. The reason for this is that addition applies in infinitely many cases, throughout the entire infinite number series. Kripke asks us to imagine someone who uses "+" as we do for all numbers up to, say, 2,000, but for numbers bigger than 2,000 he does something strange—he says, for example, that the correct answer to "What is

2,000 + 30,002?" is 7. And it's not that he has made a silly mistake in adding the two numbers; rather, all along he meant something other than addition by "+." He meant, as Kripke puts it, "quaddition," where quaddition is the same as addition up to 2,000 but then the result of quadding any number to 2,000 is always 7. This may strike us as insane, a funny form of arithmetic, but it is clearly imaginable, and nothing we had observed in our peculiar mathematician's behavior up to 2,000 could have given us any hint of how he would react after 2,000. Since addition covers infinitely many cases, no finite set of observations of someone's behavior with "+" can ever decisively settle what he means by this term. So it is not possible to *identify* someone's understanding of "+" with what he or she has done with that symbol in a finite number of cases. There is more to the understanding than that.

Here is another example: Suppose I mean by "red," not what you mean, but something that follows the rule that "red" applies to red things if they are observed before the year 2003 but to green things if observed after 2003. According to this rule, it is correct to apply "red" to red things up to 2003 but not thereafter; after 2003 it is correct to apply "red" to green things. Then I will use "red" just the way you do up to the year 2003, but I will suddenly start applying it to green things at later times. You may think I am misperceiving green things as red, but I assure you I am not; it is just that *my* word "red" has the funny disjunctive meaning just specified. Clearly, this difference in what we mean by "red" will not have shown up during my entire life history of use hitherto; it will only be revealed when we reach a time in the future when the disjunctive nature of what I mean kicks in. So my history of linguistic use does not settle what my words mean, because meaning has implications for the future. Meaning is a rule, and rules apply in indefinitely many cases.

But if understanding "+" and "red" is not a matter of my history of actual use of these symbols, then what is it? Here Kripke interprets Wittgenstein in the following way: *Nothing* constitutes a person's meaning addition by "+"; there is no *fact* about a person that makes him understand a word in one way rather than another. It is no use looking to the conscious experiences a person has when he uses the symbol "+," since these are compatible with many interpretations of the symbol; and the person's disposition to use the word in a certain way in the future is no use either, since a person may have a disposition to use the word mistakenly—and what he means is not a function of these mistaken uses. Kripke calls this Wittgenstein's "skeptical paradox"—that there is no such fact as meaning addition by a plus. But Kripke also thinks that Wittgenstein has a reply to this paradox, which he calls a "skeptical solution": instead of supposing that meaning is constituted by facts about the individuals who are said to mean things, we suppose that meaning is a matter of the way in which one person's linguistic actions compare with others in the linguistic community. We say that X means addition by "+" just when X's use of "+" agrees with ours; this is the appropriate "assertibility condition" for ascribing grasp of a rule to someone else—that their use of the rule accords with the use of others already deemed to grasp the rule. The concept of meaning is thus like the concept of fashion: It makes no sense to say that there is some fact about an individual in isolation that constitutes her being fashionable or not; being fashionable is, instead, a relational matter, involving one's conformity to others already deemed fashionable. You cannot be fashionable in splendid isolation, as a kind of Robinson Crusoe of the runway; fashionableness is essentially a communal condition, a matter of how your clothes compare to the clothes of others. Similarly, Kripke claims, a Robinson Crusoe figure cannot, according to Wittgenstein, follow a rule, because he is not a member of any community of rule followers. Indeed, he cannot have any concepts, since concepts are rules too: The rule for using the concept of addition is just that you have to give the *sum* of any pair of numbers. Thinking itself requires membership in a community. It is part of the very concept of a rule that rules are shared by many rule followers.

Now this is a very radical claim. There you are, all alone on a desert island, wide awake, your brain completely intact, and Wittgenstein says that you cannot follow rules, mean things, or even have conceptual thoughts. Granted you can't be fashionable or unfashionable, unless by reference to some distant sartorial standard. But how could the existence of other people make the difference between your being able to follow rules or not? I found this "community view" of rule-following incredible in itself, and I could hardly believe that Wittgenstein—generally regarded as a proponent of common sense—would commit himself to such an outlandish claim. And when I studied him carefully it became apparent to me that he was *not* committed to any such claim. What Wittgenstein really opposed was the idea that understanding a symbol is a matter of placing an *interpretation* on it, wherein an interpretation of a symbol is another symbol, perhaps a purely mental symbol. So when I understand "+" as meaning addition this is not a matter of supplying some *other* symbol from my mental repertoire—as it might be, an image of addition or a symbol from some supposed universal mental calculus. Nor is it a matter of any qualitative experiences I may have when using the symbol. Rather, understanding is a certain kind of ability or know-how, a *capacity* to use the symbol; and such capacities are not interpretations of the symbol but propensities to act in certain ways. Human beings have certain natural ways of acting, habits that get established by linguistic training, and these habits are the basis of meaning—not anything that "passes before one's mind" as one uses a symbol. Understanding is thus not a private conscious experience that the person knows by infallible introspection; it is a publicly observable capacity to use symbols in concrete situations. This is why you may think you understand something when you don't—because you can be mistaken about the capacities you have (compare the capacity to swim). Wittgenstein is opposing the inner and the outer, the experiential and the practical; he is not opposing the individual to the community. So long as Robinson Crusoe has the right practical skills he can follow rules as well as you or I. There is such a thing as solitary rule-following (imagine Robinson Crusoe playing solitaire to while away the time). Even if I were the only man to have ever lived I could follow rules, so long as my brain functioned properly.

I scoured the relevant Wittgenstein texts and found a great deal of evidence for this interpretation, and other commentators were also coming to the same conclusion. In fact, I thought this interpretation pretty obvious once you took the trouble to look into the matter with an open mind. I intended to write two papers on the topic—one arguing for my interpretation of Wittgenstein and against

Kripke's, the other criticizing the community view on its own merits. In the end I wrote a whole book, *Wittgenstein on Meaning*, published in 1984. I think it is fair to say that most people were persuaded, especially when other voices were added supporting the same basic position. I had certainly never intended to become a Wittgenstein expert, but I enjoyed the exegetical detective work involved, despite the fact that studying Wittgenstein was intellectually draining.

The unexpected side effect of this work, however, was to send me off in a completely different direction. I found writing *Wittgenstein on Meaning* an extremely taxing experience and felt exhausted when it was finished. Maybe this was just simple overwork or maybe it was something about the nature of Wittgenstein's writings, but I sorely needed a break from philosophy. I had been hard at it for twelve years and felt worn out, depleted. So what did I do? I wrote a novel—which I shall report upon in the next chapter, when we are finished with California.

As I observed earlier, it rained an awful lot during my second visit to L.A. I remember seeing Steven Spielberg at the height of his fame trapped by a heavy shower after a showing of *The King of Comedy* in Westwood Village. There he stood in a small crowd of other moviegoers, waiting for the rain to let up, as people whispered to each other about his celebrity presence; he looked highly uncomfortable, but obviously not enough to get his hair wet. Clearly, Hollywood and rain don't mix. For me, the continual rain meant being trapped into studying more Wittgenstein and playing more video games, neither of them particularly healthy activities. I can do this at home in England, I thought, where we know how to deal with rain (we grin and bear it). I decided not to move to Los Angeles, turning down the job offer I had received from USC. I will always remember those times, and be grateful for the hospitality and friendship of the philosophers there, as well as their intellectual stimulus, but I couldn't see myself living there indefinitely. So it was back to London and tutorials, tutorials, and more tutorials.

CHAPTER SIX

Consciousness and Cognition

UPON RETURNING, AMBIVALENTLY, TO LONDON IN 1982 I PUT IN the final spurt on my Wittgenstein book, while carrying out my usual teaching duties. Then I had to write a paper on the analysis of knowledge, on a deadline. After that I badly needed a break—I was philosophically burned out. I had always had a yen to try my hand at fiction; I had written horror stories as a boy and I had committed a fair amount of poetry to paper (I had a poem published in the school magazine back in Blackpool). Reading the early novels of Martin Amis (before he got famous) also stimulated me; I liked his combination of literacy and vulgarity, the high and the low.

One day, then, at the beginning of the long summer vacation, I started to write a novel, loosely based on the life of my brother Martin. The story concerns a young provincial art student who comes to London hoping to make it, but finds the city much harder to negotiate than he imagined. After a number of gruesome mishaps, he meets a female dentist and thereby improves his lot. My protagonist, Dave Green, is an antihero, a not very likeable young man, and the novel is comic in a bleak and miserable way. It is about loneliness in Earls Court (where I lived at the time), the fear of failure, the ordeals of sex and love—the usual type of thing. The title was *Bad Patches*, referring to both periods of time and elements of a personality. It was not at all philosophical—expressly so. I told the story in the first-person voice of my hero, an unliterate young man with loutish tendencies. Much of it is set in the seedy off-license (liquor store) in which he works with two highly unsavory, but pathetic, middle-aged men, whom he nicknames Fock and Fack, because of the way they pronounce a certain word. I wanted to avoid slipping into my habitual philosophical style, so I made the prose vernacular and slangy. I needed a break from philosophy, remember, so the last thing I wanted was to write a book about, or sounding like, a philosopher.

I completed the novel some months later, finding the experience of writing it liberating yet demanding, and wondered whether it might be publishable, though that was not my intention when I started. An agent took it on and sent it around to the London publishers, but, despite some glimmerings of interest, no one wanted to take it. Meanwhile I had started writing short stories, two of which I published in magazines, and again my agent tried to convince a publisher to take a collection of them. But again, no dice—though the response was not entirely discouraging. To this day that novel and those stories sit wanly in my bottom drawer. I have a fondness for them, and I'm glad I wrote them, even though they remain unread. I like to write, and philosophy is only one mode of writing.

This, then, is the context in which something entirely unexpected happened to me. I mentioned Gareth Evans in chapter 3, the brilliant and charismatic young Oxford philosopher whose seminar I first attended as a graduate student. Evans had been promoted to be Wilde Reader in Mental Philosophy at Oxford at the precocious age of thirty-three. This was a prestigious position with lots of research time and no undergraduate tutorials to give, and everyone felt that he deserved it. Shortly after taking up the position, however, he fell mysteriously ill. This followed an incident in Mexico, where he was a visiting professor, in which he had taken a gunshot wound to the leg in an abortive attempt by terrorists to kidnap the philosophical Mexican friend he happened to be with at the time, Hugo Margain, whose father was Mexican ambassador to the United States (Margain was fatally wounded during this attempt). Within a few weeks of falling ill

Evans was diagnosed as suffering from advanced cancer, and in a matter of months he died, in August 1980. This came as an enormous shock to the philosophical world. Evans was a person in whom the pulse of life beat strongly—a powerful squash player and avid motorcyclist, a man of boundless energy and zest. He was just thirty-four when he died. It would be hard to think of anyone less likely to end his life so prematurely. Since he was undoubtedly the best young philosopher in Britain, the tragedy of his death was more than personal. I often think of what he might have achieved by now; as it is, his writings have exercised an important influence on the state of philosophy. His posthumous book, *The Varieties of Reference*, which he worked on during his last months, and which was polished for publication by his colleague John McDowell, is still a major text in the philosophy of mind and language. His early death was probably the biggest loss to analytical philosophy in the twentieth century, rivaling that of Frank Ramsey, the Cambridge philosopher who died at age twenty-eight from jaundice in 1923.

Evans's death left the Wilde Readership vacant. This job involved interacting with the Oxford psychologists and acting as a mediating channel between psychology and philosophy; accordingly, the Wilde Reader's office was traditionally located in the Oxford psychology department. Given my background in psychology, it might have seemed that I was a potential candidate for the position. But I had been rejected several times for lowlier jobs at Oxford and I was still young myself, so I decided not to apply for the position, deeming it pointless. However, I didn't want it to appear that I had a grudge against Oxford or that I was making some sort of statement by not applying, so at the last minute I submitted a brief letter saying I wanted to apply, but supplying none of the requested items that were supposed to go with the application—curriculum vitae, latest written work, names of references. This seemed to me unnecessary because I was quite sure that I had no chance of getting the job. Everyone knew that Christopher Peacocke, my old contemporary from the B.Phil days, whom I had stopped seeing some years before, was the designated internal candidate; it was inconceivable that the job would not go to him. I didn't give the matter a second thought.

The interviews were scheduled to occur a few weeks later and I was not invited to attend—which didn't surprise me at all. But then, a week or so later, I had an unexpected call to attend for an interview. I had no idea why this was and suspected that it was just a PR move to make it appear that Oxford was at least willing to consider plausible outside candidates for its prize positions. I therefore attended the interview in a relatively relaxed mood, confident that I was not a serious candidate, but also determined to show that I knew my stuff. I was grilled by about ten people—a mixture of psychologists and philosophers—and I thought I handled myself adequately. The psychologists, in particular, seemed happy with my performance. My aim in all this was not to try to secure the position but rather to show that there were strong outside candidates who should be given a fair shot, Oxford being notoriously insular in its appointments record.

I still don't know quite what happened behind the scenes, but soon after the interview I was offered the job. I was totally astonished. Rumor has it that the psychologists found Peacocke too difficult to understand, and indeed his work can be (in my view) almost preposterously unclear. What use would he be to them if they had no idea what he was talking about? I, on the other hand, was as clear as daylight—whatever my other failings may be—and I understood where the psychologists were coming from, having been one myself. As I speculatively reconstruct it, then, it was the psychologists who prevailed over the philosophers and pushed the vote in my direction—ironically enough, given my earlier disenchantment with psychology. But, I should stress, I have no solid evidence for this supposition. In any case the job was mine for the taking, I think to everyone's surprise. After being consistently rejected by Oxford for a decade, I was suddenly offered a plum Oxford position, far superior to the other positions I had applied for in the past. And of course this greatly helped soothe the (somewhat low-level) resentment that had smoldered in me for some years. Most British philosophers have a love-hate relationship with Oxford, and I had ample cause to feel both sides of that polarity.

But I had my misgivings about accepting the job. I now liked living in London and felt apprehensive about living in Oxford, which is, after all, just a small market town sixty miles distant from the capital. I also had very mixed feelings about the Oxford philosophical scene, which I thought was in a bad phase—insular and scholastic. And I knew that it was going to be difficult living up to what would be expected of me; I was replacing Gareth Evans and I had displaced Christopher Peacocke, both quintessential Oxford figures (remember that I hadn't even been an undergraduate there). And there was one other reservation, by means inconsiderable: I was spending most of my time writing fiction, not philosophy. That was where my mind was, but I would have to be totally on the philosophical ball when I took up the Wilde Readership. I could hardly say that I was having a fallow period writing racy novels, and would probably be back to philosophy in due course. In effect, the Wilde Readership curtailed my fiction writing and forced me back to serious hardheaded philosophy. But I couldn't find it in myself to turn the job down either—for one thing, it would be the end of teaching undergraduate tutorials!

I moved to Oxford in the summer of 1985, eleven years after I had left Oxford as a graduate student. I believe I was the youngest Reader in the university when I joined the faculty. I acquired a flat in north Oxford and buckled down to the serious business of Mental Philosophy. (Brian Farrell, who had preceded Gareth Evans as Wilde Reader, once told me that his prospective mother-in-law had greeted him with the words "So you are the Mental Reader in Wilde Philosophy.") I had two major adjustments to make. First, I was to be among psychologists again after fleeing the subject thirteen years earlier. My daily colleagues at UCL had been philosophers—the people I had lunch with and talked to all day—but at Oxford my days would be spent surrounded by psychologists, and psychologists are not known for their respect for philosophy, or their understanding of it.

It was ironic to have achieved a philosophical peak, professionally speaking, only to find myself back where I started, a halfhearted psychologist. But psychology had changed a good deal for the better in the interim, and I was determined to mold myself to my new position; and it would be nice to make use of some of that psychological training on which I had spent four years at Manchester. Why, I might even do some experiments of my own. Second, I had to adjust to college life. The Wilde Readership comes with an affiliation to Corpus Christi College and I was to become a fellow of that college. This meant that I would have to attend Governing Body meetings, involve myself in college affairs, and mingle with scholars from other fields. Not much in the way of video games there, I'm afraid. But again, I was determined to do my best and to approach Oxford with an open mind. It was certainly pleasant to be the recipient of the esteem that automatically goes with being an Oxford don.

In the event, all went smoothly enough. The psychologists were gracious and welcoming, and my colleagues at Corpus were generally nice and unstuffy. I even came to enjoy having conversations with people who were not philosophers, a salutary habit I have persisted in to this day. For my first lecture as Wilde Reader I decided to put on a suit—that is, I wore a jean jacket that matched my usual Levi's. The room was packed with people who had come to see the spanking new Wilde Reader from out of town. Was I very nervous? Sure I was, but not as much as I had expected; by now I was used to this kind of thing. I set to work.

My class was entitled "Thought," which impressed me as a sufficiently resounding title. I was concerned mainly with the issue of "externalism," which connects two topics I have discussed earlier in this book. Hilary Putnam, a distinguished Harvard philosopher notorious for his abrupt changes of opinion and soaring IQ, had invented a famous thought-experiment some years earlier designed to show that "meanings ain't in the head," as he memorably put it. Here's how it goes: Imagine that the universe actually contains two planets, one called Earth and the other called Twin Earth. These two planets are remarkably similar, miraculously so. Just as there is intelligent life on Earth, so there is intelligent life on Twin Earth; moreover, the humanoid life-forms on the two planets are indistinguishable from one another. On Twin Earth there are speakers using a language just like English, which contains, in particular, the word "water." Twin Earth speakers use this word to refer to the colorless transparent liquid that fills their lakes, comes out of their faucets, and can be bought in plastic bottles with EVIAN written on them. Amazingly enough, if you look inside the heads of these folks you will see brains indistinguishable from ours, even down to the last molecule (remember this is science fiction designed to make a conceptual point, not empirical speculation). So people from Earth and Twin Earth are "molecular duplicates": They are physically indistinguishable, perfect twins; they are like identical cars coming off the assembly line, or human clones. Question: What do the Twin Earthians mean by "water"? The natural answer is that they mean the same as we do by our word "water," since their use of the word is identical to ours. But here Putnam introduces a clever twist in the story: Actually there isn't any water on Twin Earth, there is only a liquid that looks and tastes like water. In fact, their liquid has a totally different chemical formula from real water and behaves differently in experimental conditions (it will boil at 100 degrees Fahrenheit, say, unlike our liquid). Instead of being made of H_2O the Twin Earth liquid is made of XYZ, which makes it a different chemical altogether. XYZ presents the same *appearance* as water to ordinary observation, but it isn't really water (compare the way vodka looks just like plain water). So Twin Earthians refer to XYZ when they say "water," not to H_2O. Despite the fact that we and they are exactly alike in what is going on inside our heads when we use our words "water," we refer to different things on our two planets.

Why is this? Because what we refer to is a matter of our contextual and causal relations to the environment around us, not just a matter of the internal subjective states that are present in our minds when we speak of that environment. H_2O and XYZ present the same perceptual appearances, but their different distribution in the environments of the two sets of speakers ensures that the speakers make reference to different substances. If I parade a pair of identical twins in front of you, one at a time, and ask you to refer to them with the demonstrative term "that person," then you will refer to whichever individual is actually standing there, whether or not you can keep track of who is who.

This is a way of making the point that reference is not fixed by the descriptions you might offer to single out the object of reference, since there are cases in which you don't *know* any descriptions that could single the object out—after all, the twins both look and sound exactly alike. So Putnam's first conclusion is that reference is not determined merely by what is in the speaker's head. His second conclusion, a corollary to the first, is that *meaning* is not determined by what is in the head either, since meaning determines reference. We and Twin Earthians *mean* something different by the word "water": We mean *water*, but they mean another liquid, XYZ, to which we can give the name *retaw*. Hence, as Putnam puts it, "meanings ain't in the head."

It is a short step from this to the result that thoughts aren't in the head either. What do Twin Earthians think when they produce sentences containing the word "water"? They can hardly be thinking about *water* because they have never come across any water—any more than we are thinking about their liquid retaw when *we* use "water." They are thinking thoughts like: Retaw is thirst-quenching; the retaw in this lake is polluted; retaw, retaw everywhere, and not a drop to drink. But these are quite different thoughts from the ones we have, since they are thoughts about different propositions (if they think that retaw boils at 212 degrees Fahrenheit they think something false, because XYZ boils at 100 degrees Fahrenheit, as we stipulated earlier). So what you think—what concepts you have—is also partly determined by the environment in which you live and breathe and have your being; it is not just a matter of the mental and physical stuff that goes on inside your head. In a sense, then, there is more to thinking than your brain activity. Thought is essentially a brain-environment interaction or interlocking; thoughts have the content they have because of the world in which you happen to be embedded, not merely by virtue of your internal states.

I agreed with Putnam about meaning, and indeed I was the first person in print to suggest that his externalist thesis could be carried over to thought—though the point is actually quite obvious. But in my lectures at Oxford I was concerned to investigate how generally Putnam's point applies: Are *all* thoughts externally determined, or is externalism true only for a subclass of thoughts? I was also interested in the question of whether externalism could be established without the use of Twin Earth cases. Here matters turn rather technical, so I won't pursue these questions in much depth; let me just indicate how the issue works out for perceptual experience. Perceptual experiences, such as seeing a red rose, present the appearances of things; they are limited to the surface of objects, to what is manifest about them. H_2O and XYZ present, by hypothesis, the same appearance; their molecular difference is not manifest to ordinary perceptual experience of them. Accordingly, there is no difference in perceptual experience between Twin Earthians and us: The stipulated environmental difference makes no difference to how things perceptually appear to them. Nor can we imagine a case in which perceptual experiences differ while what is in the head remains the same. Therefore the content of perceptual experience *is* in the head—though, of course, objects in the environment *cause* such experiences to occur.

So Putnam-style externalism is not true for all sorts of mental states, but only for some. And I argued that we could make the same point about many other kinds of mental state—sensations like pains and tickles, thoughts about color and shape, ethical thoughts, mathematical thoughts, and others. Putnam had overgeneralized his externalism. However, this doesn't mean that these kinds of mental state are *purely* internal. According to me, though Twin Earth cases cannot be constructed for these thoughts, it is still true that their content is determined by things that lie outside the person's head—specifically, by objective properties that objects may have. If something looks square to me, then my experience has the property of squareness as part of its content, and squareness is a property that external objects have. So even for perceptual experience a more limited kind of externalism is true—what I called "weak externalism."

I eventually wrote a book on all this, *Mental Content*, published in 1989, which is a fairly technical treatment of externalism and some other issues about cognition. It was a far cry from fiction, but very Wilde Readerish—hard-core philosophy of mind with a clear relevance to psychology. At this time cognitive science was coming to the fore and I made a point of acquainting myself with this work. Cognitive science is a kind of blend of psychology, philosophy, linguistics, computer science, and neurophysiology. Its central tenet is that the mind works like a computer: Mental processes consist of symbol manipulations according to programs—word games in the brain. It is really a descendant of the Chomskian perspective I described in chapter 2. Chomsky had maintained that the mastery of language consists of an internal competence in a grammar, a system of symbolic representations. Speaking and hearing therefore involve the manipulation of these symbolic structures according to rules. The task of the psycholinguist is to figure out what these rules are and how they are implemented in the mind/brain of the speaker. There is a symbolic structure in your head, largely innately based, and understanding speech involves the activation of this structure. This is essentially similar to the way a computer implements a symbolic program in its hardware and uses it to perform various tasks. Hence the computer model of mind: Mentation is computation.

This perspective was extended with notable success to the understanding of vision, particularly by David Marr, a brilliant young British scientist working at MIT, who also tragically died around the same time as Gareth Evans (I never met Marr). Marr made a famous three-way distinction between the computational problem faced by a cognitive system, the algorithms that the mind uses to solve this problem, and the hardware in the brain that implements these algorithms. In the case of seeing an object, this three-way division breaks down as follows. First, there is the computational task of recovering from the two-dimensional image on the retina the nature of the three-dimensional object in the environment that causes this image. It turns out that this is a very difficult task, because the retinal image contains only very fragmentary information about its environmental source. So the human visual system has its work cut out for it in performing this feat of reconstruction. This stage of inquiry corresponds to Chomsky's project of characterizing the grammar of a language that a child has to acquire. The point in both cases is to gain a precise sense of what a given cognitive faculty must achieve in order to do its work. The second question is how in fact the visual system does the extracting of information from the retinal image—specifically, how it moves from the 2D image to the 3D visual percept. For example, how does it use discontinuities in light intensities to figure out that an *edge* has been detected? After all, shadows also result in such light discontinuities, but the visual system has the task of classifying these as shadows, not real physical edges. As Marr showed, this is a highly complex and difficult process, and the visual system has to function like a brilliant applied mathematician, even to see a simple object. The third stage is asking what neural structures in the brain run these algorithmic programs, and how they succeed in implementing them.

Clearly, these are all massive research problems, but Marr and his colleagues combined mathematics, computer science, and experimental results to come up with sophisticated models of how the eye works. You think seeing something is a simple task because it happens so rapidly and automatically, but beneath the surface it is as if a thousand high-powered scientists are laboring away. What you experience is the end result of an intensive assembly line of computational processes.

The relevance of this to philosophy is that it provides a

detailed theory of the kind of thing the cognitive mind is: It is a complex calculating machine, taking limited data and constructing elaborate hypotheses on the basis of these data. It is not that the brain just receives the imprint of an object and delivers this to your visual awareness neat. Rather, there is a mass of unconscious processing that goes on whenever you see anything—all those little scientists in your nervous system slaving away. As Marr pointed out, the visual system brings a wide range of built-in assumptions to the task of seeing the world, without which seeing would be impossible—assumptions about what kinds of object might be responsible for what kinds of retinal image. This leads to an enormous enrichment in our ideas about how the mind works. Our minds enable us to see objects and understand speech as if these were the simplest things in the world; nothing in our conscious awareness of what is going on gives us any clue as to the complexities of computation that underlie the simplest act of perception. If there were nothing more to the mind than our conscious awareness, then we would be unable to see anything or to process speech. There may or may not be a Freudian unconscious of repressed desires and memories, but there is an unconscious of computational processes involved in all mental acts. We might think of this as an enabling or facilitating unconscious, in contrast to the disruptive unconscious postulated by Freud.

One of the issues I was concerned with in *Mental Content* was whether these computations are always carried out in a languagelike structure. Is it that there are words and sentences that are manipulated in all mental processes, or might there be cognitive structures that *model* the world, in the sense that they mirror its denizens? Psychologists have explored this idea of mental models—isomorphisms between mental representations and objects in the world—and I investigated the philosophical credentials of the idea. The idea of mental models is particularly attractive when considering mental images. If I form an image of a red sphere this seems to involve constructing in my head a picturelike entity that shares its form with real-world red spheres; and if I imagine rotating this sphere in my mind's eye this seems analogous to actually rotating a real red sphere. So the mind does seem to work with model-like representations as well as with sentencelike ones; it's not as though when I imagine a red sphere I simply say "red sphere" to myself.

These were the kinds of issues raised by the burgeoning field of cognitive science, and as Wilde Reader I made it my business to investigate these issues. I did my job, in other words. I became a fully paid-up philosopher of psychology.

An incident stands out in my mind from this period. I was a member of a discussion group for philosophy dons that Freddie Ayer had established while he was still a professor at Oxford (he had retired before I returned to Oxford). A member of the group would read a paper to be commented on by those present (after an interval during which drinks were served). One week it was my turn and I was reading a paper about the so-called syntactic theory of mind, which says that our thoughts don't have content after all but are merely empty symbols in our head. As I read my paper I could see that Michael Dummett, whose views I discussed in chapter 4, and who was then the Wykeham Professor of Logic, was impatiently drumming his fingers on the table.

Dummett was a rotund man, with a shock of white hair stained by nicotine, who emitted an almost continuous chuckle but was prone to outbursts of temper. Someone once said that he looked as if he had been dunked in a barrel of flour and then licked his lips—so white and pasty was his appearance. When the discussion period began he rudely and violently attacked my use of the phrase "syntactic theory of mind," which he seemed unaware was current at the time. When I tried to explain it he interrupted me even more rudely and condescendingly. I had no idea what could have provoked this attack and was dumbfounded; for some minutes I lost the power of speech, and the room fell silent. Some other members of the group tried to speak, but it was clear that no more discussion was going to be possible that day. I certainly had no desire to have any further discussion with Dummett, after his rudeness, and the meeting came to an early end. Afterward all but one of the members of the group wrote me letters sympathizing with me for being attacked by a very senior member of the faculty, and asking me to rejoin the group, observing that Dummett was notorious for such irascible outbursts. But it was too late: The experience had been spoiled for me and I had no wish to enter into philosophical discussion with Dummett again. He and I had an unpleasant exchange of letters, and the group, in my absence, collectively petitioned him to apologize to me. He did so grudgingly and halfheartedly several weeks later. My suspicion was that he had been a supporter of Peacocke for the Wilde Readership—Peacocke had been his pupil and then his colleague at New College—and that he had been overridden by the others on the selection committee. This irritated him and he somehow held me to blame. In any case, it was a nasty moment, petty and ill-willed, and I cannot think of it now without a feeling of anger and injustice. Mostly people in Oxford were civilized and welcoming, but this incident stands out as an unpleasant reminder of academic ire. I believe that it is always important to maintain a polite tone in philosophical discussion, no matter how much one may disagree with the other person, and this is especially important when dealing with people junior to

THE MAKING OF A PHILOSOPHER

oneself. Trying to bully and intimidate junior colleagues is not tolerable.

But let us get back to philosophy itself, where the air is purer. I was also thinking around this time about consciousness and the mind-body problem, and what I am about to describe was the most exciting event in my mature intellectual life. I had read and reviewed Thomas Nagel's *The View from Nowhere*—an exceptionally rich and interesting book, full of philosophical substance—in which Nagel continues his earlier interest in consciousness and its relation to the physical world. Nagel's general message was that the problem of how the conscious mind is related to the brain is a profound philosophical problem. He rejected the reductionism that was orthodox in philosophy, arguing that it is not possible to give a complete analysis of what consciousness is in purely physical terms. His method of argument invokes the experiences of bats: When bats use their sense of echolocation to fly about in the dark they have sensory experiences that are very different from any we have, so that it is not possible for us to use our own types of experience as a model for theirs. We are to bats as a person blind from birth is to those with sight; in both cases there is ignorance as to what it is like to have the experiences that are missing. To know what it feels like to have a certain kind of experience you have to have had something at least similar to it yourself.

So, as Nagel famously insisted, we don't know what it's like to be a bat; bats have a form of consciousness that is conceptually alien to us. And clearly this is a general feature of consciousness: If there are conscious minds out there in other galaxies, they may be quite alien from our point of view, and not be even imaginable to us. But bat brains are different: We can see them, cut them open, investigate their neural circuitry. Bat brains are open to empirical investigation by human beings and by any intelligence with the capacity for scientific knowledge. So we *can* know about the workings of bat brains, but we *can't* know the nature of bat experience. How then— Nagel asked—could the mind just *be* the brain? How could consciousness be reducible to brain processes, given that the two are so different? More simply: If you consider your consciousness as you experience it from the inside it seems quite unlike the neural processes that someone might observe in your brain, a different type of phenomenon altogether. So how could the former be nothing but the latter?

This kind of point seems a powerful problem for materialist theories of the mind, but it is also hard to come up with any other kind of theory. Traditional dualism, which regards the mind as quite separate from the brain, also has its traditional problems, such as how the mind interacts with the brain, whether the mind can cause anything, why we need a brain at all, if the mind can get along without it. Now I had been deeply puzzled by the mind-body problem even as a psychology undergraduate. I used to wonder where mental processes were located: They had to be located somewhere, I felt, but was it an inch from the brain, was it between the neurons, was it perhaps in some other kind of space entirely? In the end I concluded that they had to be located just where their neural correlates were, and so I became a crude kind of identity theorist, identifying mental events with neural events.

But this didn't satisfy me even then, because there was the nagging feeling that this didn't do justice to the nature of mental events; it was a kind of intellectual force majeure, squeezing consciousness into the only box at hand, whether it fit or not. I was still thinking about this in the mid-eighties in Oxford, no closer to a resolution, not satisfied with the various theories currently on offer. I also happened to be thinking a lot about realism and whether reality had to be knowable by human beings. I had always believed that reality might well outstrip human knowledge, reality being one thing and human knowledge another. Nagel's example of the bats seemed a good example of this: The bats have the experiences, all right, but we humans aren't able to know what they are like. Maybe there are also things about the physical cosmos that we won't ever be able to penetrate—the origin of the universe, the ultimate structure of matter and energy. In a way, it is surprising that we know as much about the world as we do, given that we are recently evolved creatures with a finite brain capacity. If reality is independent of knowledge, as realism supposes, then it is entirely a contingent matter which facets of it are knowable to us. So two sets of thoughts were mingling in my mind at this time: the potential unknowability of reality, and the deeply puzzling nature of the mind-brain relation.

Perhaps you can guess where I am heading. One night, as I lay in bed turning these things over in my mind (I very often let my mind turn to philosophy last thing at night; it's amazing what can pop up by morning), the two sets of ideas locked together. It was one of those flashes of insight that I had read about in other people's memoirs. Maybe the reason we are having so much trouble solving the mind-body problem is that reality contains an ingredient that we cannot know. We have only a very partial grip on both mind and brain, but if we could remedy this ignorance the solution to the problem would be immediate and uncontroversial. It's like one of those detective stories in which the detective has only limited information and cannot for the life of him see how to solve the mystery—the crime looks quite impossible to explain in his current state of information—but then he lights upon the crucial missing clue and everything falls into place. But with the case of the mind-body problem, I surmised, the clue is not *going* to come to light, which explains why we have been mystified by it for centuries. It *might* come to light, I thought, but it would have to be very different from anything considered so far; it would certainly not be some minor tinkering with one of the theories currently around. And in my bones I felt that there was some deep-seated obstacle in our intellectual makeup that prevents us from discovering the missing clue.

So I set about investigating this possibility, trying to find out what it is about our intellects and the nature of mind and brain that prevents the two from meeting in a happy resolution of the problem. This at least seemed like a promising hypothesis to pursue: We are suffering from what I called "cognitive closure" with respect to the mind-body problem. Just as a dog cannot be expected to solve the problems about space and time and the speed of light that it took a brain like Einstein's to solve, so maybe the human species cannot be expected to understand how the universe contains mind and matter in combination. Isn't it really a preposterous overconfidence on our part to think that our species—so recent, so contingent, so limited in many ways—can nevertheless unlock every secret of the natural world? As Socrates always maintained, it is the wise man who knows his own ignorance.

I wrote up a version of these ideas in a paper titled "Can We Solve the Mind-Body Problem?" and sent it to the *Journal of Philosophy*, one of the premier journals in the field, which had accepted every paper I had sent them so far (even my more youthful efforts). They rejected it without explanation after several months had passed. That was when I knew I must be on to something. Originality and rejection go hand in hand when it comes to the rather staid academic journals. I didn't mind the rejection that much, but I was concerned that my paper should reach a wide audience. Fortunately, Simon Blackburn, the then editor of *Mind*, unhesitatingly accepted the paper, and it came out a few months later. It is by far the most cited, reprinted, and translated paper I have ever written. In the next chapter I shall tell the story of what it led to, but it was clear, early on, that the paper hit a nerve. Not many people agreed with it, but the position it advocated had to be added to the list of possible answers to the mind-body problem. I have now become strongly identified with the thesis of this paper and am generally known as a "mysterian" (a term introduced by Owen Flanagan). I sometimes want to protest, "I have written other things, you know."

One of my jobs as Wilde Reader was to act as examiner for the John Locke Prize every year. There were three of us, with the other two examiners rotating—I was the constant fixture. You will recall from chapter 3 what a large role the John Locke Prize played in my academic life; without it I doubt I would have been able to become a professional philosopher at all. Now I was the one deciding who got it and who didn't. I could never go through this process without experiencing a *frisson* of anxiety and relief as I recalled the days, some thirteen years earlier, when I sat dictating my illegibly written papers to a typist. I always bent over backward to be as fair as possible in this examination and tried to award the prize as often as I could, but there were years when we examiners felt that there was no one who deserved it, and we regretfully posted an empty list. I was also called upon to examine the B.Phil, which I had barely scraped onto in my student days, and again the irony was not lost on me. This examining was especially onerous because I had to read about a dozen dissertations in a term, each about a hundred pages long, on a wide variety of subjects. The Wilde Readership was proving more of a chore than I had expected. In addition I had the responsibility of supervising graduate students and at one stage had ten of these to deal with. The work of reading their stuff and meeting with them regularly to discuss it was quite intense. The idea that I would have endless amounts of free time to pursue my own research turned out to be wide of the mark (though I didn't miss those undergraduate tutorials I keep harping on).

After three years at Oxford I was entitled to a sabbatical term. I had an invitation from CUNY in New York to visit for a term. This seemed like a good plan: I liked the idea of spending some time in Manhattan, and the philosophers in New York were interesting to me, particularly Jerry Fodor and Thomas Nagel, whom I had met earlier. Oxford was feeling small and claustrophobic; I couldn't walk down the street without running into someone I knew, or who knew me, and the place lacked the ambience of a big city that I missed from my London days. Manhattan seemed like the perfect antidote. I would exchange dreaming spires for gleaming skyscrapers. So, in the winter of 1988 I turned up in New York City and was installed in a big gloomy apartment near Columbia University that belonged to Isaac Levi, a Columbia philosophy professor on leave in Oxford. It was actually in this apartment that I put the finishing touches to *Mental Content* and wrote "Can We Solve the Mind-Body Problem?" on a borrowed computer. As it transpired, this trip to the States was to prove far more momentous than I could have guessed when I arrived there on an icy January day.

What I noticed immediately—and not without a wry smile—was that my status as Wilde Reader at Oxford carried instant kudos in America. My own view was that Oxford had become a bit of a philosophical backwater, with the best philosophy being done in America; but I was happy to accept the presumption of superiority so generously heaped upon me. At the same time, I also noticed, there were some who relished the idea of knocking someone down from what they perceived to be a pedestal, and I had my fair share of this kind of criticism. When I gave an invited lecture at CUNY on the mind-body problem the usual lecture room was too small for the throng of people who turned up, and the venue had to be shifted to a big auditorium in the basement. I realized that I had "arrived," but I was unsure on what basis. Was being Wilde Reader at Oxford enough? Was I simply more popular in one country than the other?

I greatly enjoyed this visit to New York—philosophically and otherwise—and got to know many of the philosophers there, including Fodor and Nagel. When it came time to leave I felt a sense of loss, and the smaller life of Oxford seemed confining by comparison. I had to ask myself whether I could see spending the next thirty or so years as Wilde Reader in Mental Philosophy in the provincial English town of Oxford.

CHAPTER SEVEN

Metaphilosophy
and Fiction

BY THE LATE EIGHTIES PHILOSOPHY IN BRITAIN HAD BECOME A depressing business. Government policy during the Conservative Thatcher years imposed severe cuts in university funding, which meant that teaching positions that became vacant were seldom filled. This was very demoralizing for graduate students seeking to make a career out of teaching philosophy, it is hard enough completing a graduate degree in philosophy, but if you know that there will be no employment at the end of it, no matter how well you do, that is apt to take the wind out of your sails. Morale was therefore low. My meetings with my graduate students often consisted of hand-wringing and expressions of hopelessness. Pay for those in university employment was also minimal, and there was a general feeling of not being valued. Nor did things portend much better for

the future. In addition to this—and partly because of it—
the quality of philosophy being produced in England was
not conspicuously high. Oxford, in particular, had fallen
into an insular phase in which its own rather feeble products were touted as great insights. The distinguished
philosopher Elizabeth Anscombe, an authentically terri-
fying woman, friend of Wittgenstein, now dead, is credited
with the remark that there is nothing worse than hearing
one second-rate Oxford philosopher telling another second-
rate Oxford philosopher how brilliant he is—and that
about summed it up for me.

On many occasions I would try to introduce ideas from
across the Atlantic into philosophical discussions only to
be greeted with incomprehension and a mild sense that I
had just committed a faux pas. Meanwhile students would
be treated to obscure, narrow, empty disquisitions on
what some other Oxford philosopher had said last week in
his or her seminar. I didn't care for this intellectual atmosphere, to put it mildly. It meant, among other things, that
I had little opportunity for the sort of open-minded, rigorous, and informed discussion that is so important to a
philosopher. Too often the conversations I did have were
with my colleagues in the psychology department, and
they tended to veer toward the question of what the point
of philosophy was anyway. Psychologists have a tin ear for
philosophical issues, and their training imbues them with

the conviction—I might say prejudice—that the only real questions are the kind that can be settled by means of controlled experiments; abstract conceptual questions simply don't compute with the majority of them. I tired quickly of this kind of lunchtime discussion, in which the value of my discipline was routinely questioned by people without the faintest idea of what it involves. I was often reduced to advising my interlocutor to *read a book*.

Soon after I returned from New York to Oxford I had a job offer from Rutgers University in New Jersey. Now the name of Rutgers does not, for most people, quite have the resonance that Oxford commands. But around this time Rutgers was establishing itself as a major center for philosophy, making a number of senior appointments of first-rate people. Their chief coup was hiring Jerry Fodor and attaching him to the cognitive science department as well as the philosophy department. Fodor (who is now a close friend) is a gentle man inside a burly body, and prone to an even burlier style of arguing. He is shy and voluble at the same time, a cat lover and a philosopher slayer. He is a formidable polemicist burdened with a sensitive soul. He likes to refute his opponents into an early grave in the afternoon and then quiver at the opera in the evening. Disagreeing with Jerry on a philosophical issue, especially one dear to his heart, can be a chastening experience; even when he is most wrong he seems to be winning the argument. His quickness of mind, inventiveness, and sharp wit are not to be tangled with before your first cup of coffee in the morning. Well, adding Jerry Fodor to the faculty at Rutgers instantly put it on the map, Fodor being by common consent the leading philosopher of mind in the world today. I had met him in England in the seventies and spent a good amount of time with him in New York during my sabbatical there. I found him to be the genuine article, intellectually speaking (though we do not always see eye to eye). Rutgers was now poised to build on this foundation; inviting me to join the department was part of this effort. But I was unsure: Moving to the USA was a big step, not least in terms of the practical issues that had to be faced. I have never been one for practical issues, finding them at best distracting and at worst totally paralyzing. What would I do with my furniture? Did I really have to fill out all those immigration forms? What if I came and didn't like it after all?

I asked Oxford if they would give me a term of unpaid leave so that I could visit at Rutgers to see how I liked it there (this kind of dispensation had been allowed to others before). They denied my application, to my surprise, and the late timing of the decision meant that I was already committed to going to Rutgers. The General Board, which made the decision—a committee not made up of philosophers—refused to relent or to explain their verdict, even after I pointed out the serious financial, and other, problems their refusal would cause me, and they would not meet with me to discuss the matter. In the end I took the term of leave unilaterally, which got me into even more trouble with the General Board, and I was compelled to resign my position at Oxford without properly trying out Rutgers and without obtaining a green card. This left a bitter taste in my mouth and caused me much anxiety and inconvenience. The philosophers at Oxford were on my side in all this, but the faceless members of the General Board were determined not to let me take a term's unpaid leave, which, among other things, would have saved Oxford some money in lean times. I still have no idea what their reasoning might have been. None of this, needless to say, sat well with me. I felt the sting of arbitrary authority, which evidently was indifferent to the welfare of someone with no power and a great deal at stake. I resolved at the time of my resignation not to do anything to help Oxford University in the future—this being the only feeble protest I could make—and I have stuck to that resolution. So my time at Oxford ended unpleasantly, abruptly, and dramatically, after only three years in the job. I have not been back since. Emotionally, it is over between Oxford and me, despite our occasional good times together and my youthful infatuation with the place.

I shall spare my patient reader the details of my practical trials in making the move to America. Suffice it to say that they were traumatic and soul-destroying (fellow practical-phobes will understand me). Getting any philosophy done was out of the question. This was particularly trying because I was in the middle of my work on consciousness and cognitive closure, and I wanted to get this work written up and published. I was living in Manhattan by now, on a short-term visa, and commuting to Rutgers, trying at least to do my teaching, while I dealt with the Immigration and Naturalization Service and other worthy institutions. But finally this searing and wearing phase came to a conclusion after a year or so, when I obtained "resident alien" status, and I could get back to work. I wrote three more papers on consciousness, developing the position I had set out in "Can We Solve the Mind-Body Problem?" I put these together with some older work of mine on the mind-body problem and published a volume entitled *The Problem of Consciousness* in 1991. I had no high hopes for this book, as it was really just a collection of papers mostly already published, but it must have been the right moment, because the book sold extremely well for an academic text (which is not to say very well) and had to be reprinted several times.

Consciousness was in the air, and later years would see a blossoming of interest in the topic. I suspect that my principled pessimism about solving the problem acted as a spur to some people to try to come up with something. Daniel Dennett, in particular, published a book called *Consciousness Explained* soon after my book appeared, saying that it couldn't be. My thesis seemed to enrage the volatile Dennett, and he even wrote, in a review of my book, that he was "embarrassed to be in the same profession" as I for having dared to suggest that the problem of consciousness might not be solvable by the human intellect. I thought this remark was beyond the pale, and my hitherto friendly relations with him went downhill from there. When I subsequently presented a lengthy criticism of Dennett's book at a meeting of the American Philosophical Association, with Dennett present with the oportunity to reply, the atmosphere was markedly unfriendly, and he later accused me of trying to "get him back" for his review of my book. In fact, I simply found his position implausible and his arguments unconvincing—and said so.

I date the publication of *The Problem of Consciousness* as the time at which I came out of the academic closet and made my descent from the ivory tower (if you will forgive the mixed metaphors). My ideas about consciousness started to attract interest from outside the narrowly academic philosophical world. *Scientific American* ran a story about consciousness in which I was featured, along with a photograph of me looking like Anthony Hopkins in *Silence of the Lambs* (it is often said that there is a resemblance between the actor and myself); beneath the picture I was described as a "hard-core mysterian." *Time* magazine had a cover story about conscious minds and computers, prompted by the chess match between Deep Blue and Kasparov—in which I was quoted in the same paragraph as none other than Shirley MacLaine, well-known mystic and reincarnationist. Japanese *Newsweek* ran its own feature in which I was photographed cradling a conical light for eerie effect. *Omni* ran a story too, with no special effects. I also appeared briefly on *Nightline* with Ted Koppel, talking about minds and machines, and in a Canadian documentary about consciousness. I even appeared in *Art Forum,* of all places, accompanied by a picture of the old rock band The New Mysterians, after whom my position had been named. This was all good fun and made me realize that philosophy could be interesting to people not professionally engaged in it.

At the same time three other extracurricular activities occupied me. First, I began writing more for magazines and journals that did not have an exclusively philosophical readership: the *Times Literary Supplement*, the *London Review of Books*, the *New Republic*, *Lingua Franca*, and others. I even wrote an article on gyms for *New York* magazine, since I had become a bit of a gym rat myself. True, I was writing mainly about philosophy books, but I was at least addressing myself to a wider audience and learning how to convey difficult concepts to people not professionally trained in them. I had always valued clarity in writing, and this was an opportunity to bring that clarity to people who might be put off by the obscure jargon habitually used (or abused) by academic philosophers. I believe that philosophy can be of interest to everyone, even the kind of "dry" philosophy I have been presenting in this book. I want to give people the real thing, and I want them to be engaged by it.

My second new direction involved a foray into practical ethics. Since moving to America I had been troubled by the quality of public debate about practical ethical issues, such as drugs, abortion, violence, and so on (not that things were any better in England). It struck me that the calm, measured rationality of philosophy might usefully be brought to bear on these issues, and that what was really needed was a book that could be read by a typical high-school student or beginning undergraduate. I was not aiming to make any original contributions to these subjects but simply to bring them together in a clear and accessible form. My emphasis was on the kinds of practical decisions that confront real people about sex, censorship, self-defense, eating meat, having an abortion, taking drugs.

I suppose my general position would be described as

liberal, but I am a staunch antirelativist in ethics and believe strongly in the idea of virtue and right conduct; it is just that my idea of right conduct may not coincide with that of the Moral Majority. I favor tolerance, individual freedom, and kindness to others. Above all, I favor thinking clearly and open-mindedly about the issues. So, for example, I believe that clear, unclouded thought recommends that drugs should be decriminalized, both because of considerations of personal freedom and as a matter of sensible public policy. The book I wrote, entitled *Moral Literacy, or How to Do the Right Thing*, was published in 1993 and has been used by colleges quite widely. Recently I had an E-mail from a student at Columbine high school who had read the book and wanted to ask me a question about violence. This kind of writing for the general public does nothing to gain respect from one's philosophical colleagues ("Oh, McGinn has gone totally pop lately, and he used to be an okay philosopher in his own way"), but to me popular writing is extremely important, and I wish I could be more effective with it. I admire Peter Singer's work in this direction, though I cannot write about practical ethics with the same single-minded conviction that he can muster. It is, as they say, dirty work, this writing about morality, but somebody has to do it.

My third departure from the narrow academic path goes back to fiction. My first novel, *Bad Patches*, had not found a publisher, and maybe that was no bad thing (though I do have a soft spot for the book), but I was not yet washed up as a novelist. While I was in the process of leaving the country in which I was born and moving to America—undergoing all the wrenching experiences that go with that—I conceived the idea of a novel that would deal with this kind of dislocation, but in a more thematically psychological way than is typical. My story concerned an ordinary insurance salesman, Alan Swift (specialty: cars, collisions), living in Holloway in London, who one day decides to take a trip to New York for a temporary escape from his confining life, not telling his wife and young son where he is going—he just up and disappears. Once in New York he overstays his visa, unable to bring himself to return to England, and becomes an illegal alien, working in a photocopying shop and remaking himself in various ways. His superficially prosaic story is interlaced with two other narrative strands: One corresponds to his imaginative fantasy life, which withers away as he escapes his past; the other strand corresponds to a voice from his unconscious, which vacillates uneasily between claustrophobia and agoraphobia. This latter is the voice of fear, anxiety, and dread. The whole is a meditation on travel and home, and is entitled *The Space Trap*. At home we feel the constrictions of claustrophobia, but if we stray too far from home, agoraphobia becomes our condition.

The novel is a kind of horror-comedy in which space itself is the protean monster, ever-present and uncontrollable.

This time I did find a publisher for the book, and it was published by Duckworth in England in 1993, along with *Moral Literacy*. The redoubtable Colin Haycraft was my editor, with whom I struck up a lively friendship; sadly, however, he died only a couple of years later. The novel received some favorable reviews, notably in the London *Times*, but never really took off; it didn't make the grade to paperback and is now out of print. Still, I did succeed in writing a novel that was actually published, and some people do seem to have enjoyed reading it. However, I didn't feel that the book was enough of a success to justify my spending much more time writing fiction, especially when I had philosophy to write—and I knew I could do *that* adequately enough. One day I may get back to writing fiction, but I haven't constructed a piece of dialogue in nearly ten years. I occasionally feel the itch, but there is always something more pressing to do.

My philosophical work around this time involved turning back on the nature of philosophy itself, perhaps appropriately, since I had now been a philosopher for twenty years. The philosophy of philosophy is called "metaphilosophy": It inquires into the nature of philosophical problems, the possibility of philosophical knowledge, what methods to adopt in order to make philosophical progress. It is perhaps the most undeveloped part of philosophy, and hence fraught with controversy. One of the perennial questions of metaphilosophy is why philosophy does not make the kind of steady, assured progress that we associate with the sciences and with history, archaeology, even literary scholarship. In philosophy it seems that every generation repudiates the supposed insights of the previous generation, so that there is no cumulative body of philosophical knowledge that everyone can agree to; philosophers always seem to be bickering and dithering, to put it unkindly. Thomas Kuhn talked about periods in science of revolutionary progress, on the one hand, in which "paradigms" are dramatically overthrown, and normal science, on the other hand, in which scientists work steadily to increase knowledge within a common set of assumptions. In philosophy we never seem to get the normal phase, in which a body of theory is generally accepted and built upon; everything always seems more or less up in the air.

To be sure, every few decades someone thinks of a new method or approach that is promised to yield the kind of solid advance other disciplines enjoy, but it usually doesn't take long before this is torn down and discord returns like a cantankerous relative. And it isn't as if philosophers are a bunch of dimwits, relatively speaking, like fourth-graders trying to figure out astrophysics for themselves. I venture to suggest that philosophers tend on the whole to be persons of considerable intelligence, many of them highly competent at science, and endowed with excellent thinking skills. It's not that if you let some real scientists loose on philosophical problems they would have all the answers for you in a matter of days. In fact, when scientists, particularly distinguished ones, try their hand at philosophy—usually after they have retired—the results are often quite inept, risibly so. So what *is* it that makes philosophy so hard? Why do we still have no proof that there is an external world or that there are minds other than our own? Why is freedom of the will still so hotly debated? Why do we have so much trouble figuring out what kind of thing the self is? Why is the relation between consciousness and the brain so exasperatingly hard to pin down?

There have been answers to these metaphilosophical questions, favored by different philosophers at different periods; there has been no shortage of metaphilosophies, each as hotly contested as the issues they purport to illuminate. The traditional view, stemming from Plato, is that philosophy is concerned with an especially ethereal and remote region of reality—the world of abstract Forms or Universals that we only glimpse dimly. To know what beauty really is, for example, we would need to have a clear apprehension of the Form of beauty; but this Form lies in a region of reality—often referred to as "Platonic heaven"—that we cannot gain clear-sighted access to in our mortal state. The questions of philosophy are thus simply more profound—more elevated—than the questions of science, which is concerned merely with the ordinary empirical world of sensory observation. Some parts of reality are just deeper than other parts, more complex or subtle. Much of philosophy is concerned with what Plato would have called the soul, and the soul is the deepest of realities, the murkiest of pools, the subtlest of substances. Philosophy is the study of The Profound.

Another view, at the other extreme—and popular during the middle part of the twentieth century—is that philosophy consists of a bunch of meaningless pseudo-questions. Given that these questions make no sense, they obviously have no satisfactory answers. Philosophical questions are like the "question" of why colorless green ideas sleep furiously, or how tall no one is, or what becomes of a number if you immerse it in cold water. These are just not proper, well-formed questions; they merely have the grammatical form of questions. This type of metaphilosophy was advocated in an unusually sharp form by the Logical Positivists, a group of philosophically minded scientists working mainly in Vienna in the thirties. They divided meaningful sentences into two classes: Those that could be verified by means of observation and experiment—

such as the sentences of chemistry and physics—and those that are mere tautologies, such as "bachelors are unmarried males" and "everything is what it is." Since the questions that philosophers debate—such as "Is the will free?" or "Is the mind a separate substance from the body?"—fall into neither category, they must be meaningless, and should therefore be banished from respectable intellectual discourse. No more philosophy for you! Philosophy is just portentous-sounding gabble.

Wittgenstein and the so-called Ordinary Language philosophers of the postwar period adopted a somewhat softer version of essentially the same idea. They held that philosophical problems arise through misunderstanding our own language and then using it in ways that it cannot sustain. Thus we never normally say "the will is free" or "human actions are determined by law and causality" in ordinary speech, so these sentences are ipso facto under suspicion of meaninglessness. And when we stick with ordinary examples of sentences about human action we find that there is no problem about their truth: "Johnny decided to go to the store to buy some food; no one prevented him; and he returned safely by bus." Where is the philosophical problem in that description of what went on? And isn't that all we mean by cumbersome philosophical talk about "free will"? We just need to attend to our actual vocabulary for talking about human actions, and we will see that no insoluble problems are raised by it. The problems arise when we invent a new philosophical vocabulary or distort words we use in one context by transferring them to another context. The philosopher needs therapy, not solutions—something that will cure him of his professional tendency to gorge on misused language.

A third view, popular toward the end of the twentieth century, is that philosophy is just immature science. Here the idea is that what we *now* call philosophy is just the residue of problems left over as science has eaten up more and more of what *used* to be called philosophy. First, mathematics and physics spun off from what was called philosophy; then biology made its own way in the world; psychology is the most recent spin-off as it works to establish itself as a solid science. So philosophy *does* make progress; it gradually gets transformed into science. It only *looks* as though it doesn't because we use "philosophy" as a label for what hasn't spun off yet. Philosophy of language will turn into linguistics, philosophy of mind will turn into psychology, metaphysics will eventually become a branch of regular physics. "Philosophy" is simply the name we use for subjects that haven't grown up yet.

I myself think that none of these views is right. It seems to me that standard philosophical problems are perfectly meaningful, are not incipient science, and yet are not concerned with some remote and subtle region of reality. I take my cue from my position on consciousness: The reason we cannot make solid progress with the mind-body problem is that our human intelligence is not cut out for the job. Perhaps, then, that is the explanation of philosophical intractability more broadly; philosophical problems are of a kind that does not suit the particular way we form knowledge of the world. The question then is what it is about the problems and our intelligence that makes the latter unsuited to the former. More generally, what is the broad structure of human intelligence, and what determines its strengths and weaknesses? My ideas on these questions are contained in my book *Problems in Philosophy: The Limits of Enquiry*, published in 1996. I won't go into this in detail here, since my recent book *The Mysterious Flame* (1999) already contains a comprehensive popular treatment of these ideas. But I will give the flavor of the metaphilosophy I advocate (in case you don't already own that book).

In working on the question of the scope and limits of human intelligence I rediscovered Noam Chomsky, who had been such a large influence on me in my undergraduate years. Chomsky had long championed the idea that what we call human intelligence is really a collection of "cognitive modules" that are specialized in their area of application. Knowledge of language, for example, is a separate module with an innate basis that fixes its scope and mode of operation; this module enables us to acquire human languages with remarkable rapidity, but it is not equipped to acquire languages with radically different kinds of grammar. In general, what we vaguely call human intelligence is a matter of specific systems that are richly structured and dedicated to particular tasks; intelligence is not some sort of infinitely plastic all-purpose knowledge-acquiring device. There are "intelligences," not some uniform, overarching faculty of intelligence. Reading Chomsky again made me see the relevance of this conception of human intelligence to metaphilosophy; philosophy is hard in the way it is for the same sort of reason human children would have trouble learning Martian language—the task and the tool are not made for one another. It's like trying to crack nuts with a feather duster.

I started to correspond with Chomsky on these questions, and it was nice to be collaborating with someone who had been such an intellectual hero of mine twenty-five years earlier. He even kindly read the entire manuscript of *Problems in Philosophy* and wrote me detailed comments on it (he did this while on a plane traveling to somewhere in South America to lecture; Chomsky's intellectual speed and capacity are legendary—he seems to be the only human of whom his own theories of cognitive limitation are not true!). The central conjecture of my book is that there is a certain cognitive structure that shapes our knowledge of the world, and that this structure is inappropriate when it comes to key philosophical problems. I call this the CALM conjecture, short for Combinatorial Atomism with Lawlike Mappings. Roughly speaking, you understand something when you know what parts it has and how they are put together, as well as how the whole changes over time; then you have rendered the phenomenon in question—CALM. Natural entities are basically complex systems of interacting parts that evolve over time as a result of various causal influences. This is obviously true of inanimate physical objects, which are spatial complexes made of molecules and atoms and quarks, and subject to the physical forces of nature. But it is also true of biological organisms, in which now the parts include kidneys, hearts, lungs, and the cells that compose these. The same abstract architecture applies to language also: Sentences are complexes of simpler elements (words and phrases) put together according to grammatical rules. Mathematical entities such as triangles, equations, and numbers are also complexes decomposable into simpler elements. In all these cases we can appropriately bring to bear the CALM method of thinking: We conceptualize the entities in question by resolving them into parts and articulating their mode of arrangement.

Understanding is a matter of decomposition and recombination, seeing how nature "hangs together," what its anatomy looks like. In this way we come to grasp how one kind of entity depends upon others: The physical objects need the atoms, the organisms cannot exist without the organs and cells, the sentences derive from the words, the three-dimensional figure is constructed from planes, lines, and angles. Nature is a system of derived entities, the basic going to construct the less basic; and understanding nature is figuring out how the derivation goes. The CALM structure is the general format for this kind of understanding: Find the atoms and the laws of combination and evolution, and then derive the myriad of complex objects you find in nature. If incomprehension is a state of anxiety or chaos, then CALM is what brings calm. Question: Does CALM work in philosophy?

Consider the way the mind depends upon the brain. The brain itself is a complex spatial object, made up of tiny cells called neurons, connected by fibers called dendrites and axons. Consciousness evidently depends upon the activities of these interacting cells: A particular group of cells fires and you experience, say, a specific shade of red. Your experience is somehow derived from these neural activities, which are electrochemical in nature. But the mode of derivation involved does not fit the CALM structure; it is not that your experience is literally *com-*

posed of the corresponding neural activities. If your experience is composed of anything, it is at the phenomenal level—for example, you might have a complex sensation of a spherical shape with a dent in the side, the whole colored green and red in an irregular paisley pattern. But of course you do not, in the typical case, have a sensation *of* neurons and their cellular activities (you would have to look into your brain for that). The brain activities themselves have parts that fit the CALM structure—smaller cellular processes, molecular reactions, etc. But the experience, as it is apparent to you while having it, does not resolve itself into such more elementary physical processes. If it did, there would scarcely *be* a mind-body problem, since we would be able to grasp how exactly it is that the brain gives rise to conscious processes.

The whole problem is that the conscious mind is not something that emerges from the brain as a whole emerges from its parts. Consciousness is not, on its face, a thick wedge of brain tissue, with phenomenal parts corresponding to the parts of the wedge. Somehow the brain generates the mind, but it does not do so by means of simple spatial aggregation. So the mind is quite unlike regular organs of the body, which are simply aggregations of more elementary biological parts. This is, of course, precisely why the dualist is convinced that the mind is something quite separate from the brain.

In *Problems in Philosophy* I argue that something very similar holds in the case of the self, free will, meaning, and knowledge. There are yawning gaps between these phenomena and the more basic phenomena they proceed from, so that we cannot apply the CALM format to bring sense to what we observe. The essence of a philosophical problem is the unexplained leap, the step from one thing to another without any conception of the bridge that supports the step. For example, a free decision involves a transition from a set of beliefs and desires to a particular choice; but this choice is not *dictated* by what precedes it—hence it seems like an unmediated leap. The choice, that is, cannot be accounted for simply in terms of the beliefs and desires that form the input to it, just as conscious states cannot be accounted for in terms of the neural processes they emanate from. In both cases we seem to be presented with something radically novel, issuing from nowhere, as if a new act of creation were necessary to bring it into being. And this is the mark of our lack of understanding. The existence of animal life *seems* like an eruption from nowhere (or an act of God) until we understand the process of evolution by natural selection; we can then begin to see how the transitions operate, from the simple to the more complex. But in philosophy we typically lack the right kind of explanatory theory, and hence find ourselves deeply puzzled by how the world is working.

This message is not very congenial to the optimistic philosopher who wants to *solve* the deep problems that brought him or her to philosophy. For I am saying that this is a futile aim; my book could equally have been called *The Futility of Philosophy* (in fact, I wanted to call it *The Hardness of Philosophy*, but my publisher told me it would never sell with a title like that). How (as they say) does this make me feel? Both good and bad. Good, because now I don't have to blame myself for not making more progress in solving the problems that trouble me— any more than I blame myself for not being able to leap to the moon. What is not humanly possible cannot be a source for self-rebuke. Bad, because it would have been nice to arrive at the definitively correct solution of these problems—something that would take the puzzlement away. Philosophy must now be admitted to be a condition of terminal puzzlement, a permanent fretting ignorance— and there is little joy in that. Mathematicians struggled for centuries to prove Fermat's last theorem, finding it peculiarly recalcitrant, knocking their heads against a brick wall. Then one of them, through hard work and ingenuity, actually came up with the required proof, and everyone agreed that it was right. Problem solved! It would be wonderful if something similar could be done with a longstanding philosophical problem—the mind-body problem or the problem of free will. Some ingenious individual finally figures out the correct solution, by hard work and ingenuity, and everyone else agrees that it is indeed correct. But if I am right, this isn't going to happen. That is rather depressing. And my philosophical colleagues are not thrilled by my pessimistic position either, though some concede that it *may* be right. The aim of a university, they think, is to advance knowledge, and I am saying that in this case that aim is futile.

But I don't think they should be quite as shocked as they are: Didn't we already know that human faculties have their limits? We don't know what it's like to be a bat, after all, and there would be futility in a lecture course that purported to impart this knowledge to sophomores (Bat Experience 101), just as there would be futility in a lecture course that offered to teach blind people what it is like (from the inside) to see colors. Gydel's theorem famously shows the limits of provability in a formal system: There are truths of arithmetic that no formal system can generate. Nor does anyone think that we can ever discover how many blades of grass covered Henry VIII's lawn, or how many individual dinosaurs there were. There are grammatical sentences in English that cannot be parsed by any human speaker because of excessive demands on our speech-recognition capacities. Reality can obviously outstrip our ability to know about it in all sorts of ways. Is it really all that surprising if the most frustrating area of human intellectual endeavor—the sub-ject we call "philosophy"—should also contain questions we cannot in answer in principle? On reflection, isn't the existence of such humanly unanswerable questions quite *predictable*? We aren't gods, after all; we are recently evolved organisms made of pretty low-tech materials.

So, after twenty-five years of philosophy, have I argued myself out of a job? Well, not exactly: I have just redefined the job. Maybe we can't *solve* the main problems of phi-losophy, but we can at least reflect on them, formulate them clearly, spell out the various options, develop a sense of their depth. As Bertrand Russell pointed out in *The Problems of Philosophy* in 1914, the value of philoso-phy does not lie in the acquisition of what he called "pos-itive knowledge"—as the value of science does—but rather in enlarging the imaginative scope of our minds and in appreciating that ignorance is part of the human condition. Human life can never be perfect, and human knowledge is no different. Besides, as I shall explain in the next chapter, there is still a lot of philosophy that can be done, even when it is agreed that some is beyond us.

New York suited me, and so did Rutgers. In Oxford when the day's work is done there is nothing much to do except visit the bookstore; it's a very quiet town, especially during vacation. But in New York there is always something inter-esting to do, even if it's just walking down the street. You may think that the bustle of New York must be a distraction to someone whose main occupation is writing and thinking. But I find the opposite to be true; philosophy can easily become obsessive, so that you spend your life locked up in your own circling thoughts. You can become "sicklied o'er with the pale cast of thought," as Shakespeare remarked of the overly pensive Hamlet. And this state of mind is not conducive to good philosophy, as well as being a downer. Philosophy is something you need to be able to stop doing or it will devour you. Manhattan, for me, is the perfect way to stop doing philosophy; it provides an escape from the obsession, a rude jolt of teeming life. Every day, when I am working in my study, I gaze out of the window at Broadway and know that there is a writhing human world out there waiting to engulf me when I step outside. This is comforting, and it helps me keep a sense of proportion. Oddly enough, I find it salutary that the world outside doesn't care that much about philosophy—whereas in Oxford it seemed to be the be-all and end-all. I can keep philosophy at a safe distance if I am vividly aware that other people hardly even know what it is.

Rutgers continued to augment its philosophy department, hiring several excellent people after me. The intellectual atmosphere there has suited me very well, with its directness, clarity, and honesty—as well as high philosophical talent. I found that I enjoyed teaching undergraduates again, which I had ceased to do as Wilde Reader at Oxford (and no tutorials!). Bringing philosophy to young minds full of energy and curiosity has its own rewards, even if those minds are not always as sophisticated as one might ideally (and unrealistically) wish. I try never to forget my own student self: keen, receptive, but hardly well-honed. I keep my radar on for that lone student, with the haunted look about the eyes, the hesitant question, the expression of relief when understanding dawns—that exemplary student who might someday want to become a philosopher.

During the 1990s the New York area was in the process of establishing itself as the world center of philosophy. With Princeton and Rutgers in New Jersey, and Columbia, CUNY, and NYU in New York City, the general area now has the strongest field of philosophers on the planet, and the level of philosophical activity is high in every sense of the term. Moreover, philosophers from all over the world regularly pass through New York. There is almost *too* much good philosophy going on. So my move away from Oxford has meant an enhancement in intellectual stimulation, and I can't see myself ever leaving New York. It was a long journey from Blackpool to Manhattan, but it seems to have ended happily.

CHAPTER EIGHT

Evil, Beauty, and Logic

WHAT WAS I TO DO, HAVING DECIDED THAT THE PROBLEMS IN philosophy that most interested me could not be solved? Retire? Keep explaining to other people why their philosophical ambitions are futile? Find something to work on that offered more hope of progress? By chance, one of our course offerings at Rutgers had no teacher for the year— Philosophical Problems in Literature. I volunteered to teach it, though I had never taught anything like this before. My intention, only vaguely articulated, was to use works of literature as materials for a course on ethics. My own interest in fiction writing was one motive for this, but I also liked the idea of turning to something completely new; so at age forty-three I branched out into alien territory (midlife intellectual crisis?). I chose four novels that I particularly admired and which had interrelated themes, centering on beauty and evil: *Lolita,* by Vladimir Nabokov; *The Picture of Dorian Gray,* by Oscar Wilde; *Billy Budd,* by Herman Melville; and *Frankenstein,* by Mary Shelley. My plan was to elicit the moral themes of these works and to evaluate the characters that appear in them. Close reading of the text was to be my method, guided by whatever I knew of real life—this was not to be a scholarly or historical course in the usual Eng Lit style. I wanted to engage with the material, and the students, at street level, so to speak—at the level of ordinary human experience. This was to be a course on the neglected topic of *life*. As befits that aleatory topic, I presupposed little in the way of methodology, just patience, sympathy, and a sense of right and wrong.

The course turned out to be a success, despite an initial feeling among the students that they didn't know what we were up to, and I have now taught it several times. I always start with *Lolita*, a widely misunderstood book with a strong (albeit strange) moral texture. Naturally, this has required me to read the book several times, and I can quote large parts of it from memory now; it lives in my brain, a perpetual presence, a second voice. I have also had occasion to discuss the mechanics and ethics of the novel on numerous occasions. When Adrian Lyne's film version of *Lolita* was causing some controversy in 1998—it could not find a distributor in America—I wrote an essay defending the morality of the novel and published it in the *Times Literary Supplement*. This was translated for a German newspaper and quoted in *The New York Times*, which led to my being asked to appear on CNN to debate the morality of the book with a critic of it. What I mainly remember of this debate, which was broadcast live, was that the cameraman neglected to tell me where to look while I was talking, and his wild gestures of direction in mid-utterance were most off-putting. Still, I think I made the right points, explaining that by the end of the novel Humbert Humbert's pedophilia has received a sharp and appropriate punishment, and that even he (ladies and gentlemen of the jury!) recognizes the depth of his maltreatment of Lolita.

I have to admit, however, that discussing the topic of pedophilia on live television is a fairly daunting undertaking, especially when you are defending a book written in the seductive voice of a confessed pedophile. The things we do for art. Actually, I think the film, as opposed to the novel, is far too easy on Humbert Humbert: He is a much nastier character than Jeremy Irons represents him as being, and much less sane. The novel is, indeed, a work of great literary genius, and the voice of the narrator creepily beguiling, but Nabokov never leaves us in serious doubt about the evil of his protagonist, at least if we read the novel with care and sensitivity. He is a murderer, a rapist, a liar, and a madman. On a purely personal level, studying the novel is a humbling experience for anyone who has tried their hand at fiction: Nabokov's talent is so awe-inspiring that it makes you want never to put clumsy pen to paper again.

The main result of teaching this course, so far as my own work was concerned, was that I wrote a book entitled *Ethics, Evil and Fiction* (1997). One of the chapters is entitled "The Evil Character," and in it I attempt to delineate and explain what the psychology of an evil person is like. This too earned me a couple of TV appearances, in which I discussed various psychopaths and evildoers. Characters in the novels I was teaching were a stimulus to this work, but I was also prompted to consider the evil character by some events in the news, particularly the murder of the child James Bulger by two young Liverpool boys. My general theme was that evil characters invert the usual sympathy we feel for the sufferings of others: Instead of suffering when others suffer, they revel in the suffering of others. The sadist is precisely someone who derives pleasure from the pain of others. I distinguished the motives of the pure sadist from those of someone who is instrumentally evil: A person may cause suffering to others in the course of achieving some other goal, as with violent theft, but a sadist is someone who causes pain in others for no further reason.

This is what we think of as pure malevolence. The character of James Claggart in *Billy Budd* is purely evil, because he seeks to destroy innocent Billy for no reason other than to destroy him; he derives no benefit, such as promotion, from the hoped-for destruction. I suggested that deep existential envy plays a large role in this kind of evil—particularly envy of virtue and innocence. Claggart wanted Billy dead because of his envious feeling that he could never attain to Billy's goodness of nature; hatred of goodness stems from envy of it. I think moral philosophers should discuss this kind of topic more, but they typically have little to say about it; perhaps the lack of rigorous methodology deters them. In any case, literature clearly provides a rich field for moral reflection. It is no accident that we speak of "the moral of the story."

The other main theme of my book is beauty. In *The Picture of Dorian Gray* beauty is set in opposition to morality, as Dorian supernaturally retains his exceptional good looks while his portrait takes on the signs of age and sin. An obsession with beauty leads Dorian into sin, as he tries to make his life into a work of art, even its evil parts. He elevates beauty above virtue, in short, with disastrous consequences. In *Frankenstein* the scientist's monstrous creation exemplifies the reverse of Dorian Gray—he is ugly on the outside, virtuous within (at least until the bitterness of rejection turns him violent and vengeful). Both stories emphasize the essential duality of human beings: their inner spiritual being and their outer corporeal appearance.

We tend to be taken in by appearance, entranced by the beautiful, repelled by the ugly. But a person's inner self can be radically at variance with their appearance, and is not so easily accessible. In my book I argued that *Frankenstein* is a kind of exaggerated parable of normal human existence, in which we are identified with the monster: We too have bodies that are not in every way aesthetically pleasing (especially under the skin); we too suffer the pangs of unfair rejection; we too are born without our consent into a hostile world, with death as the only escape. We constantly sense the duality of our outward appearance and our inner being, as we negotiate human society (think, for example, of skin color). This duality is the root of human self-consciousness, and much human misery.

I also argued that the ugliness of Dorian Gray's soul illustrates what I called the "aesthetic theory of virtue": Virtue may well not coincide with outer beauty, but it is identifiable with inner beauty—beauty of soul or character. I devoted a chapter to this ancient Platonic theory of virtue, which is not much in favor with today's analytical moral philosophers, arguing that it was a plausible and illuminating way to understand moral virtue. Dorian sacrifices his inner beauty to his outer, giving off an illusion of virtue, but inwardly he is vicious and corrupt, as his repulsive picture attests. He devotes himself to beauty, seeking aesthetic value even in evil, but the irony is that he creates for himself a soul whose ugliness is indelible. Thus beauty and virtue are not so separate after all; there is a harmony between them at a fundamental level.

Evidently, *Ethics, Evil and Fiction* is a very different kind of philosophical work from the kind I have been talking about earlier in this book. It is closer to human concerns, more akin to drama and literature and film, less rigorous in formulation. Some reviewers of the book called me "eccentric" for writing it, meaning (I suppose) that it was not orthodox. But part of my point was that moral philosophy today fails to address itself to issues that literature and other imaginative forms deal with, and that we need to incorporate those forms into our database for moral reflection. The parable and story are often used to impart moral understanding; it is not just a matter of memorizing a list of moral imperatives, like the Ten Commandments, which promote obedience rather than understanding. In my view, then, there needs to be a marriage between ethics and literature, for the benefit of both. (If I were teaching ethics in high school I would do it by means of literature, not religion.)

Meanwhile, during the nineties, what I think of as my second career heated up. For years I had been writing reviews for such journals as the *London Review of Books* and the *Times Literary Supplement*, but now I started writing more frequently for the *New Republic*, *The New*

York Times, the online magazine *Slate*, and more recently for the *New York Review of Books*. I am reputed to be a tough reviewer, and it is true that I see no point in pulling my punches. Naturally this gets me into a lot of trouble with the people I review negatively. Without entering into the gory details of particular cases, I wish to make one observation about these (often unpleasant) skirmishes. When I review person A negatively person B will often praise me for my penetration and courage, remarking, "It's about time someone said that about A." But when I later review person B, employing the same critical standards as before, B will instantly assume that I have some ulterior motive, that I am willfully aggressive, that I have unconscious hatreds that I let loose in my critical reviews. B will accordingly declare me Public Enemy Number One. In fact I simply apply the same standards across the board, with the possible exception that I adopt a softer tone for younger, less established writers. What is true is that I often criticize the works of powerful, well-connected people, and this is deemed somehow unacceptable (it is certainly not particularly good for me). I don't do this, of course, because they are powerful and well-connected; I do it for the obvious reason that I think their views are flawed in various ways. Somehow people find this difficult to believe, in a way *I* find difficult to believe. But there is no point in writing reviews at all unless one has the freedom to express one's opinion forcefully and clearly. And I am on occasion just as positive as I can be negative.

Let me focus on one example of this kind of writing, so that I can give the flavor of the thing. I began writing for the *New York Review of Books* a couple of years ago, starting with a review about two books on consciousness and the brain. Then the editor, Robert Silvers, asked me if I was interested in writing an essay evaluating Freud, in connection with an exhibition devoted to Freud showing in New York. I hesitated, saying that I was no Freud expert and had no worked-out view on his theories. (You may recall that I was much stimulated by Freud as a teenager and that it was Freud's writings that determined my applying to study psychology at university.) The editor replied that this was what he was looking for—a fresh, unbiased evaluation of Freud from someone without an entrenched position; and I was interested in revisiting Freud and finding out what I thought about him now.

It seemed like an intriguing assignment. So I spent a summer reading Freud, along with some commentaries on him, and deciding what I thought. In the end I came to the view that most of Freud was ingenious fantasy, unsupported by factual evidence, dubious in itself, and often lacking in theoretical coherence. For example, Freud held that all dreams are wish-fulfillments, but it is quite obvious that we have straightforward anxiety dreams that are not the expression of any secret wish. I recently dreamed that my car got towed for being illegally parked and that I would have to go to jail and pay an enormous fine; there is nothing wish-fulfilling about that! Freud also famously held that little boys suffer from castration anxiety because of parental threats, but there is just no good evidence that this is true. His idea of repression as a mechanism that forces unwelcome memories into the unconscious faces the obvious objection that some of our most painful memories won't go away. I wrote the piece, it was carefully edited, and it was published in the fall of 1999.

Many letters of protest came in to the offices of the *Review*, most of them hostile and personal, though some were supportive. It was assumed by many that somehow this was a put-up job, that the editor and I had conspired to produce a piece that "knocked" Freud. In fact, I simply applied the same critical standards of argument and evidence that I would apply to any other book that I was reviewing, without being deterred by the fact that Freud is a cultural icon with a devoted following and an entire industry—psychoanalysis—built upon his (shaky) foundations. The main lesson for me in carrying out this work was the contrast between my earlier youthful self, not yet critically developed, ready to take things on trust, and my mature hypercritical self, primed to spot a hole in any argument, no matter who is giving it. Two long letters criticizing

my essay were published by the *Review*, in which the writers were given ample space to express their grievances; I
gave my replies at similar length. This seemed eminently
fair to all parties, and a fine example of open, vigorous
debate. Intellectual progress is fired by strong, honest criticism, not by mealy-mouthed adherence to the party line.
But this is obvious, isn't it?

As an interesting, if somewhat chilling, sequel to this
episode, let me report the following story. One day I came
home from Rutgers to find a fax from the *Review* forward-
ing me a letter about my Freud essay. Nothing unusual in
that; for some weeks I received such faxes daily. But this
one was handwritten, quite long, and very neat. Skimming
the letter, it struck me as coherent and quite sensible, all
about psychoanalysis being a pseudo-science that is
believed for the same sorts of reasons people believe in
other pseudo-sciences—as substitutes for religion. Turning
to the address at the head of the letter, I saw that it came
from a "Federal Penitentiary, Maximum Security." How
interesting, I thought, prisoners read the *New York Review
of Books*. Then I noticed the identity of the sender:
Theodore Kaczynski, otherwise known as the Unabomber.
The small, neat handwriting was that of a barely sane, but
evidently intelligent, murderer, whose slight professorial
figure I had frequently seen on television. After an initial
shudder, I thought: Well, I am certainly expanding my

readership. (Joyce Carol Oates also sent me a congratula-
tory postcard about the piece, so it wasn't all murderers
and enraged psychoanalysts.)

Speaking of expanding my readership, I hope I won't be
accused of excessive name-dropping if I mention Jonathan Miller and Oliver Sacks. I had known of Jonathan
Miller for many years from English television and radio;
his polymathic brilliance—both artistic and scientific—
astonishing memory, subversive wit, and force of personality had always impressed me. Then one day, while I was
still Wilde Reader in Oxford, I had a letter from him out of
the blue: He had read my (very demanding) book *Mental
Content* while on vacation in Greece and wanted to con-
gratulate me on it, and to give me some references he
thought I might be interested in. By sheer chance he was
scheduled to give a lecture in the Oxford psychology
department the next week, so I made a point of introduc-
ing myself to him. He seemed taken aback by my youth,
and possibly also by my size (he is very tall, I am very
short), but we arranged to meet in London in a couple of
days so that he could give me a monograph he thought I'd
like to read. This was in 1990. We have been firm friends
ever since. I would like to be able to say that he is as nice
as he is intelligent, but no one could be quite *that* nice. I
can only say that I enjoy our meetings enormously and
value my friendship with him very highly. Our conversa-

tions range all the way from philosophy of mind and cognitive science to Peter Cook and Dudley Moore routines. He doesn't share my sporting enthusiasms (about which more soon)—but then no friendship is perfect, is it? (I did once entice him to play a game of pool with me, which he approached purely mathematically.) On one occasion we were having dinner together in a New York restaurant with Jerry Fodor, when none other than Placido Domingo walked in, with whom Jonathan had worked before in directing some opera or other. Knowing that Jerry is an opera fanatic, Jonathan was careful to introduce him to Mr. Domingo. It was the one occasion on which I have seen Jerry speechless and quaking, moved, and overawed.

I met Oliver Sacks through Jonathan; their friendship goes back to their school days in north London. Compared to Jonathan, Oliver is taciturn, diffident-seeming, distracted. But he too is remarkably erudite, equally at home in chemistry and literature, neurology and music. He is also, unlike Jonathan, a sportsman: power-lifting in his younger days, now swimming and scuba diving. He is an exceptionally thoughtful conversationalist, who weighs his words as if they are precious stones, unworldly, boyish, burly—like no one else I have ever met. His intellectual passions are contagious, and his ability to bring subjects like metallurgy alive is remarkable (he has a sensual relationship with minerals). On one occasion he introduced me to Robin Williams, who played him in the film version of *Awakenings*, Sacks's book about the curative effects of L-dopa. The actor was as unstoppable as ever, giving me a truly astonishing imitation of an Indian taxi driver from London who was really a nuclear physicist. But Mr. Williams also struck me as a logically minded and very direct man; there is order beneath all that creative chaos. We exchanged notes on what led me into philosophy and him into stand-up comedy, against all expectations (he was studying political science at college).

Now that I am dropping the names of movie stars, I cannot evade the obligation of telling my Jennifer Aniston story (don't worry, we will soon be back to logic). I happened to be at a premiere party for the movie *Meet Joe Black*, courtesy of a friend in the movie business. As the crowds poured in after seeing the movie, I found myself being pushed toward the back, close to Jennifer Aniston, star of *Friends* and face of a thousand magazine covers (her now-husband, Brad Pitt, was off to the left, safely behind a rope, looking preternaturally handsome). We smiled at each other and I told her that I didn't really belong there, being a philosopher, not a movie person. She flashed me one of her famous smiles and said perkily, "Oh, who is your favorite philosopher?" Surmising that she must have taken a philosophy course in college and had some knowledge of the subject, I replied, "Probably Bertrand Russell." She looked a little crestfallen and said, "I haven't heard of him."

Sensing that I needed to wax more popular, I ventured, "Kant is also a favorite." She replied, dishearteningly, "Haven't heard of him." This wasn't going well, so I hastily resorted to "How about Plato?" "Oh yes, I know Plato," she replied, happily. Encouraged, and wanting to confirm further knowledge on her part, so that I wouldn't seem like the censorious professor and she the delinquent student, I attempted "And there's always Descartes, of course." "Haven't heard of him," she murmured, after a heavy pause. Desperately I blurted out, "Well, you are wonderful in *Friends*!" She said "Thank you" and beamed, but the damage was done. I don't know who was more embarrassed, her or me. I made my polite farewell as she threaded her way over to the glamorous Brad group. Afterward I reflected that she had probably meant *eastern* philosophy, a subject that Mr. Pitt evidently has some interest in—and about which I am about as ignorant as she was of Western philosophy. It's not, I hasten to add, that I think it somehow disgraceful of her not to know the names of more philosophers—she is an actor, after all, not a scholar. In fact, throughout our brief conversation she was gracious and pleasant, and I was no one to her. I just wish she had known who Descartes was, that's all—it would have eased the interpersonal discomfort. It's not often that Hollywood meets analytical philosophy, and it would have been nice for it to have gone more swimmingly.

Later that same evening I also had a brief conversation with Anthony Hopkins, also starring in the film. As I remarked earlier, it is often said that I resemble him, so I took the opportunity of asking his opinion. He looked me straight in the face for a long moment. People milled all around us, as he considered the question, and then delivered his judgment: "No, we don't, not really, though I see why people might think so." I agreed with him, saying that it had never occurred to me until people started mentioning it. Then he added, after a pregnant pause, with a voice that only he can command: "Well, perhaps around the eyes." With our gazes locked, I nodded my assent, the hair on my neck lifting slightly. So now, whenever anyone says I look like Anthony Hopkins, I can inform them that I have discussed the matter with Sir Anthony himself at some length and that we agree that it isn't really so— except perhaps around the eyes.

Okay, that's it for name-dropping and star-gazing; now, before we return to philosophy proper, we have to discuss the serious business of water sports. My excuse for bringing this subject up is that a book on the life of a philosopher ought to have something to say on the philosophy of life. The life of a philosopher is obviously highly cerebral, detached from the ordinary world and from nature. It consists largely of sitting alone in a room, thinking, writing. I have found it important, in maintaining a sense of balance, not to let my life become too exclusively intellectual. Playing video games was part of that, but it is not exactly the healthiest or most rewarding of activities. I have tried squash and tennis too, and pool and darts and ten-pin bowling.

I believe that everyone should engage in some form of physical activity they enjoy (and not just something they do to keep fit), for the health of both mind and body. A few years ago I discovered water sports, late in life (perhaps I was slow to come to the water because I was very nearly drowned at the beach when I was four years old—I still remember the struggling and gulping and the sense of impending death). First it was body-boarding in Florida, lying prone on a piece of styrofoam and riding in with the waves. But soon it was kayaking and regular surfing, and then windsurfing and sailing. My favorite activity of all is surf kayaking, which I do as often as possible these days (during most of the writing of this book I have alternated sitting at my computer and sitting on a kayak in the surf off the beaches of Long Island). In surf kayaking you paddle out in a short, specially designed boat with a flat bottom and thigh grips to hold you in; when a suitable wave approaches you paddle like the devil to catch the break, then surf in, maneuvering the boat as you ride the wave. The waves can be powerful and the ride is not exactly serene; you certainly can't think about philosophy when you are sliding at speed down the face of a big steep wave and fighting to keep ahead of the curl. All too often the wave will turn you over and churn you in the water with the greatest of ease, as your boat shoots away from you toward the shore. But the sensation of catching a perfect wave, controlling the boat as you skim along the line, keeping your balance as you dismount from the wave still upright, is about as exhilarating as anything I know (of course, some things are more *pleasurable*). It is the perfect counterpoint for the life of still contemplation, abstract reflection, and brain fatigue. For me it provides the kind of muscular physical activity I relished in my days as a gymnast and pole vaulter, and which as an academic I lost for many years.

Not to sound too mystical, but I also find it immensely enjoyable to be out on the ocean, at one with nature, feeling its impersonal power, negotiating what it throws my way. A good wave is like a gift from the natural world, though a double-edged gift. Even when a breaker zaps you under the roiling water there is a feeling of natural harmony, as your body regains its control. I also derive great satisfaction from the simple engineering that goes into a kayak and paddle: such a simple machine, but so perfectly suited to its job. Man, machine, and nature in tight harmony. And of course there is the pleasure of being with friends with the same passion, and teaching others the skills involved. This, then, is my philosophy of life—or at least part of life: Get a boat and ride the waves. Or anything else that unites you with your body and with nature (skiing is also good in this respect). The danger of being a philosopher is that you become all brain; you need an antidote to this over-cerebralization. Who wants to be a stalk with a clump of gray matter at the top?

My work has recently focused on philosophical logic, one of the driest and most abstract parts of philosophy. Perhaps because for some years my efforts were directed at subjects that are inherently messy and problematic—consciousness, metaphilosophy, evil, beauty of soul—I have enjoyed working on a subject that is sharp, rigorous, and pure. When I left Oxford in 1974 to become a lecturer, philosophical logic was my primary interest, so I have returned twenty-five years later to an enthusiasm of my youth, after many a divagation. In the nature of the case this logical work is harder to make accessible than the other subjects I have taken up in this book, but I shall do my best to convey the gist of it, focusing on the topic of existence (in the book that resulted from this work, *Logical Properties* [published in 2000], I also deal with identity, necessity, predication, and truth).

Existence is a peculiar matter. Some things have it and some don't. But even to say this sounds paradoxical—what are those *things* that don't exist? It would be wrong to say that they are ideas, like the idea of Sherlock Holmes, since the *idea* of Sherlock Holmes does exist—it is just Sherlock himself, the man, not the idea of the man, that doesn't exist. Just as a picture of a unicorn cannot be identified with the unicorn it depicts—since the picture of the unicorn exists but not what is pictured—so the idea of a nonexistent thing is not the same as that nonexistent thing itself. So we seem to be saying, of certain nonmental entities, that they lack existence: How is that possible? At the same time, when we attribute existence to an object that does exist we seem to be attributing something trivial and oddly impalpable. Everything exists, we want to say—that is, everything that exists exists. But that seems like an excessively common property for things to have. With other properties some things have them and some don't, as with being red or a man or a prime number. But existence seems not to distinguish one object from another. Once we know we are talking about something real we already know it exists; it is then trivial to add that the thing in question *exists*. So statements of existence seem either redundant or contradictory, depending on whether they are true or false.

Reflections like these have led some philosophers to deny that existence is a real property. In ordinary language we use the word "exists" in the same way grammatically that we use "red" and "man," as the grammatical predicate of sentences. But these philosophers hold that this is logically misleading, so that ordinary language is enticing us into the logical error of taking existence to be a property that things can either have or lack. Instead, it is maintained, the concept of existence functions like the concept of being numerous. You cannot sensibly say of a particular object that it is numerous; when you use the word "numerous" you are always speaking of some property or attribute that is being said to have many instances. If I say "the mosquitoes are numerous" I am not saying of any particular mosquito that *it* is numerous; I am saying that the attribute of being a mosquito has a great many instances. In the same way, it is suggested, if I say that Bill Clinton exists I am not saying of a particular man that he has this peculiar property of existence (which Sherlock Holmes lacks); rather, I am saying of some attribute—say, the attribute of being a president of the United States who was once governor of Arkansas—that it has just one instance. When I say that tigers exist I am saying that tigerhood has instances, not that individual tigers have the mysterious property of existence, and the same goes for cases in which I seem to be attributing existence to an individual: I am really speaking of an attribute and saying that it has a unique instance. Statements of existence are never strictly about individuals but always about the attributes of individuals—to the effect that they have instances. There is no such thing as the strange property of existence, after all.

That is the orthodox view of existence. I think it is wrong. My view is that existence really is what it appears to be—a genuine property of objects. True, it is universal to all objects that exist, but it is equally the case that blueness is a property of all blue objects. And there *are* objects that don't exist, objects of thought like Sherlock Holmes—just as there are objects that are not blue. Existence is like self-identity, a logical property that is also very widespread. The trouble with saying that existence is really a matter of an attribute having instances is that the idea of having instances already contains the concept of existence. What does it mean to say that "tiger" has instances? It means that there are things that exist which are tigers, but this is to use the concept of existence as a property of some objects and not others. The fact is that we do sometimes refer to things—like fictional characters—that don't exist; so the concept of existence does distinguish among the things we refer to. It is therefore not a trivial concept; it can be informative to be told that something exists (as when you are told that Sherlock Holmes was actually a real detective in Victorian London whom Arthur Conan Doyle chose to use as his model). Some properties just are highly general, such as the property of being an object, or the property of being thought about, or the property of existing in space and time. These are indeed more general than redness and manhood, but they are not unreal because of that.

Among other things, this shows that the traditional response to the ontological argument for the existence of God (which I discussed in chapter 1), namely that it wrongly presupposes that existence is a property,   is mistaken: Existence *is* a property of objects; the question is really whether this property can be argued to be entailed by the very concept of God—which I doubt. When I was young I was interested in the ontological argument because of its theological implications; now I am interested in it from a purely theoretical point of view. Still, it has stayed with me all these years, as tantalizing as ever. Is this why philosophers sometimes feel as if they are being haunted? Or do I mean persecuted?

This is just a taste of the topic of existence, which is obviously at a high level of abstraction. It is a very characteristic kind of philosophical question. We speak of existence all the time and it is obviously a fundamental concept, but what exactly does it mean? It proves remarkably difficult to say. The question is puzzling, but it is possible to make progress with it (unlike the problems about the mind that I earlier suggested totally resist solution). And existence is not a topic that science could expect to deal with. It is a purely philosophical question, simple but surprisingly confusing. Thinking about it makes you realize that even our most basic concepts are not clear to us; we use them smoothly enough, but we have no articulate understanding of what they involve. That is where philosophy steps in. And this shows that it is wrong to think that all genuine questions are scientific or empirical. Indeed, science itself inevitably raises philosophical questions.

The same is true of literature, history, economics, computer science, mathematics, and so on. In mathematics, for example, there is the question of where numbers come from: Are they just marks on paper, or ideas in mathematicians' minds, or are they, as Plato thought, objective mind-independent entities that exist outside space and time? Nothing you learn in a regular mathematics class will equip you to answer such questions (though your mathematics teacher may have his or her own philosophical views on these questions). In empirical science, theories are devised to explain the data that have been observed, and these theories are often supposed to provide true descriptions of reality. But notice that this banal characterization of science uses a number of concepts that cry out for elucidation: What is a theory? What is an explanation? What distinguishes an observation from the theory employed to explain it? What is truth? What is reality? Science operates with these concepts, but it is not equipped to account for them. It is the same with the social sciences: They also use the concepts just mentioned, but they also invoke concepts such as reason or motive, as well as normative concepts such as right or obligation—and these bring us to moral and political philosophy, as well as philosophy of mind. The arts employ aesthetic concepts such as beauty and representation, and philosophical questions are raised by these concepts: Is beauty subjective or objective? Is all artistic representation of fundamentally the same kind? What determines the aesthetic value of a work of art? Then there are the extremely general concepts that crop up everywhere—time, causality, necessity, existence, object, property, identity. No scientific discipline can tell you what these concepts involve, because they are presupposed by any such discipline; we need philosophy to understand these concepts. For example, is causality just a matter of the mere constant conjunction of events—of "one damn thing after another," as A. J. Ayer used to put it—or does it involve an element of necessary connection? And what kind of necessity might this be? Is it anything like the necessary truth of "bachelors are not married"?

These are all questions human beings naturally ask and which they have been puzzling over since articulate thought was first recorded. Children spontaneously ask philosophical questions, much to the frustration of their parents—since the parents are often as philosophically clueless as their children. The philosopher is just someone with a particularly strong interest in these age-old universal questions; she is the embodiment of one kind of human curiosity—the kind that seeks the general, not the particular, the abstract, not the concrete. Of course it is easy to be impatient with such questions, because they do not admit of scientific resolution. But really this response is just philistinism combined with science fetishism. Science is no doubt a fine and noble enterprise, but it is not the only valuable form of intellectual inquiry. We should not run away with the idea that a question is either scientific or nothing.

There are many excellent books that try to make science intelligible to the layperson, many of which I have read with great interest; yet very few books try to do the same for philosophy. That is what I have attempted here, by describing what it is like to be a philosopher from the inside. I hope you have gained an impression of what a philosophical life is like, at least the life of one philosopher; and I hope even more that philosophy now strikes you as a fascinating and rewarding subject for study and thought.

I AM WRITING THIS IN MASTIC BEACH, LONG ISLAND. IT IS THE long summer vacation. Outside the sun shines, the mosquitoes lurk, and the water beckons. I am thinking of a novel that has just been published in England by my friend Edward St. Aubyn, whom I met at a conference on consciousness in Arizona a few years ago. The novel is about a man who has only six months to live and who is obsessed with the nature of consciousness and how it is related to the body. He talks to various philosophers and scientists, hoping to resolve his questions, but none satisfies him. Toward the end of the novel he shares a train ride with a philosopher named McGinn who explains to him why consciousness is such an enigma: because of the inherent cognitive limitations of human beings. Consciousness is an entirely natural phenomenon, this character McGinn insists, joined to the brain by natural bonds; it is just that we do not have the right mental organ with which to understand it. The man finds this position convincing; he understands why the most obvious thing in the world—his own consciousness—should be so hard to come to grips with. Fortunately, as it turns out, there has been a misdiagnosis, and he returns to normal life. My appearance in this novel seems to sum up my life to date: It fuses my interest in philosophy with my dabbling in fiction; it expresses my desire to be a teacher who can bring illumination into other people's minds; and it captures my feeling that I am a character in a novel—but a true one, the "novel" I have composed in writing this book.

Now I think I shall go out for a paddle.